NOTES

ON

THE STRENGTH OF MATERIALS

AND

THE STABILITY OF STRUCTURES.

BY

W. M. GILLESPIE, LL. D.

PRINTED FOR THE CIVIL ENGINEERING CLASSES IN UNION COLLEGE.

SCHENECTADY, N. Y.
WISEMAN, RUSSELL & Co., PRINTERS & STEREOTYPERS.
1870.

WILLIAM J. DAVIS,
ENGRAVER.

JOHN E. RUSSELL,
TYPOGRAPHER.

ENGINEERING STATICS,

OR

STRENGTH AND STABILITY.

---

1. This treats of the strength of materials and the stability of structures, used in Engineering and Architecture; as frames, walls, roofs, arches, bridges, &c., &c.

It is divided thus:

Part I. The forces which are to be resisted.

Part II. The strength of cohesion, or the resistance of materials to breaking.

Part III. The stability of position, or the resistance of bodies and structures to overturning.

Part IV. The stability of friction, or the resistance of bodies to sliding.

PART I.

2. The forces which are to be resisted are the forces which tend to produce motion and so to destroy equilibrium. These may be statical or dynamical.

Statical Forces.

3. *Table.*

Material.	Weight in lbs. per cubic foot.
Air	.08
Water	62.50
Sand	95.
Common mould	90.
Sandy loam	100.
Clay	110.
Gravelly sand	110.
Gravelly clay	120.
Land gravel	125.
Water gravel	145.
Sandstone	155.
Limestone	160.
Granite	170.
Gneiss	155.
Brick	90 to 135.
Masonry of brick	100.
Masonry of sandstone	130.
Masonry of granite or limestone	145.
Pine	30 to 54.
Hemlock	40 to 48.
Oak	40 to 60.
Cast iron	450.
Bar iron	480.

A bar of iron one inch square and one yard long will weigh ten pounds; one bushel of wheat sixty pounds; one bushel of rye or corn fifty-six pounds; one bushel of oats thirty-six pounds; one barrel of flour one hundred and ninety-six pounds, net, or two hundred and twelve, gross; one barrel of hydraulic cement three hundred pounds; of hay eight to twelve cubic yards make one ton gross; piled wood .56 of its solid weight; a crowd of men weighs seventy pounds per square foot of surface covered by them, each averaging one hundred and forty pounds and covering two square feet. An ordinary crowd weighs from fifty to sixty pounds per square foot. When so closely packed that they cannot move they weigh ninety pounds. A crowd of men is the greatest possible strain that can come upon a common road bridge. A drove of cattle weighs forty pounds per square foot of surface covered, or about half as much as men. A double row of the heaviest loaded wagons, used on the road, with teams, weighs six hundred pounds per running foot, or about fifty pounds per square foot.

A railroad freight train averages half a ton, gross, per running foot. A train of the heaviest locomotives, with tenders, weighs one ton, gross, per running foot. Empty freight cars average fifteen thousand pounds each, and the load is about the same. Passenger cars weigh nineteen thousand pounds. Baggage cars fourteen thousand pounds. Thirteen passengers, with their baggage, average one ton, gross.

Roofs weigh as follows: A frame truss averages 5.2 pounds per square foot of ground plan. The roof, beams and plank five pounds per square foot of superficial surface; slating five to nine pounds; shingles one and a half to three pounds; zinc one and a quarter to two pounds; lead four to seven pounds; copper one to one and a half pounds; ceiling nine pounds; tin five-eighths to one and a quarter pounds.

Fluid Pressure.

4. This equals the product of the surface pressed, multiplied by the depth of its centre of gravity, and multiplied by the weight of the fluid per unit of measure. The *centre of pressure* is the point where a single pressure would exactly counterbalance all the pressures acting in a contrary direction. On a vertical rectangular surface it is at one-third the height from the bottom. The pressure of the atmosphere is about fifteen pounds per square inch.

Semi-fluids.

5. This name is applied to loose earth, grain, shot, &c. A B C D is a cross section of a wall. B E is the natural slope which the earth would take if free. Let a equal the angle E B C which this line makes with the vertical, and φ the angle it makes with the horizon. There will be another line B F somewhere between B E and B C, such that the triangular prism of earth, whose section is F B C, shall exercise the maximum thrust against the surface B C. Let x = angle C B F which this line makes with the vertical.

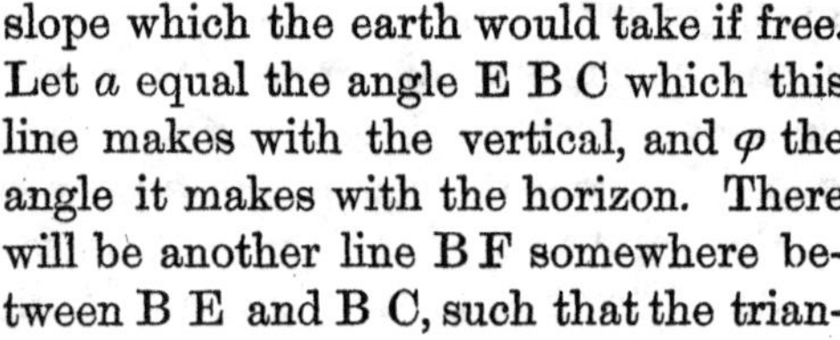

We have now to find the weight and thrust of this prism. We will consider one foot in length of it.

Let w = weight of a cubic foot of earth; then the weight of one foot of the prism equals $\frac{1}{2}$ B C $\times$ C F $\times$ 1 $\times$ w = $\frac{1}{2}$ B C $\times$ B C $\times$; tan. $x \times w$ = $\frac{1}{2}$ B C^2 tan. $x \times w$.

To find the horizontal thrust: Let G be its centre of gravity, G H = weight of prism, G I = the unknown horizontal thrust, and the resultant force acting on the surface B F equal to G K. We shall then have, horizontal thrust = G I = wt. × tan. H G K.

Now when a body is just about to slide on a surface, the angle which the direction of the force tending to produce the motion, makes with the perpendicular to the surface, equals the angle of repose for that surface, i. e., $= \varphi$.

Therefore, drawing G L perpendicular to the surface B F, we have, K G L $= \varphi$; and L G H = B F C = $(90° - x)$ ∴ H G K = L G H − L G K = $90° - x - \varphi = 90° - (x + \varphi)$.

The horizontal thrust = weight of prism, multiplied by tan. H G K = wt. × tan. $[90° - (x + \varphi)]$.

$$= \text{wt.} \times \text{cotan.}\ (x + \varphi).$$

$$= \tfrac{1}{2}\overline{BC}^2 \times w \times \text{tan.}\ x \times \text{cot.}\ (x + \varphi).$$

This is a maximum when the factors containing x are a maximum. Now tan. x × cot. $(x + \varphi)$ =

$$\frac{\text{sin.}\ (2x + \varphi) - \text{sin.}\ \varphi}{\text{sin.}\ (2x + \varphi) + \text{sin.}\ \varphi} = 1 - \frac{2\ \text{sin.}\ \varphi}{\text{sin.}\ (2x + \varphi) + \text{sin.}\ \varphi}.$$

This is a maximum when sin. $(2x + \varphi)$ is a maximum, (i. e. = 1 and then $2x + \varphi = 90$, or $x = \tfrac{1}{2}(90 - \varphi) = \tfrac{1}{2}\alpha$); i. e., the line B F bisects the angle α, which the natural slope of the earth makes with the vertical.

$$\text{Cot.}\ (x + \varphi) = \text{cot.}\ (90° - \tfrac{1}{2}\alpha) = \text{tan.}\ \tfrac{1}{2}\alpha.$$

Then the maximum horizontal thrust = $\tfrac{1}{2}\overline{BC}^2 \times w \times$ tan. $\tfrac{1}{2}\alpha$ × tan. $\tfrac{1}{2}\alpha = \tfrac{1}{2}\overline{BC}^2 \times$, tan.$^2\ \tfrac{1}{2}\alpha \times w$.

The pressure of a perfect fluid would be = B C × 1 × $\tfrac{1}{2}$ B C × wt. of fluid . = $\tfrac{1}{2}\overline{BC}^2$ × wt. of fluid.

Consequently the pressure of a semi-fluid, as earth, &c.,

against a surface, equals the pressure of a fluid whose weight equals the weight of the semi-fluid, multiplied by the square of the tan. of $\frac{1}{2}$ the angle which the natural slope of the earth makes with the vertical. This force acts horizontally against the back of the wall at the centre of pressure, i. e. $\frac{1}{3}$ the height from the bottom. It is, therefore, equivalent to $\frac{1}{6}\ \overline{BC}^2 \times (\tan.\ \frac{1}{2}\alpha)^2 \times w$; acting against the top of the wall.

If earth be exposed to saturation with water, it will act like a fluid of its actual weight, i. e. one and a half to two times that of water. The thrust may be resisted not only by a wall, but by timbers, or piling, or struts, &c.; and hence needs to be calculated independently.

When the earth is higher than the top of the wall, we may use its entire height in the formula for the pressure, when the height is not greater than twice the height of the wall.

MATERIAL.	Natural slope of earth.	Tan. of its angle with the vertical.	Weight of cubic ft. in lbs.	Wt. of cubic ft. multiplied by $\tan.^2 \frac{1}{2} a$.
Fine dry sand.........	30°	.577	94	31
Common dry earth....	43°	.435	94	18
Common moist earth...	54	.325	106	11
Very compact earth.....	55	.315	125	12.5
Loose broken stone....	50°	.364	81	11

HEAT.

6. For 1° Fah. wrought iron expands $\frac{7}{1000000}$ of its length, and cast iron $\frac{6}{1000000}$, with a force equal to about 15,000 pounds per square inch of cross section, or from six to nine tons.

Freezing of Water.

7. The force of this is very great. It has burst a bomb-shell which could resist a pressure of 18,000 pounds per square inch.

Animal Force.

8. The average force of a man drawing horizontally is seventy pounds, for a short time, and thrusting horizontally thirty pounds. A horse is equal to about six men, or about four hundred and twenty pounds at a dead pull. A nominal "horse-power" is raising 33,000 pounds one foot high in one minute, which is equivalent to raising one hundred and fifty pounds, at the rate of two and a half miles per hour; eight hours of such labor is a day's work.

DYNAMICAL FORCES.

Solids in Motion.

9. The effects of weights passing over surfaces at high speed are greater than at rest. The most important case is that of railroad trains passing over iron bridges. It is greatest when the bridge is short and therefore light. The dynamical deflection produced by a weight in motion varies as the square of the velocity, and inversely as the square of the length of the bridge. The extra stress is caused by the centrifugal force. In an experiment upon a light model the dynamical deflection was double the statical depression; this is in an extreme case. For a bridge twenty feet long the dynamical deflection may be half more than the statical, while in a bridge fifty feet long at fifty miles per hour it was only one-quarter more.

The quantity of motion in a body is the product of its mass by its velocity $= m\ v$.

This is often called its *momentum*.

The living force or "*vis viva*" of a body equals its mass × by the square of its velocity $= m\ v^2$. Some call it "living power," and half of this product "living force."

The work done by a moving body equals its mass or force multiplied into the distance through which the body moves, or force acts.

The product of a weight in pounds, multiplied by the distance in feet through which it falls, gives its work in "*foot-pounds*." Other forces are expressed similarly. The accumulated work in a body is also called its "*potential energy*."

The "moment of inertia" equals $\Sigma .\ r^2 .\ m$. (See Jackson's Dynamics, Art. 45).

Impact.

10. Impact cannot be compared with pressure, i. e., the force with which a moving body strikes another can not be theoretically expressed in the units of the weight of the moving body.

The effects of an impact and a pressure may however be compared in certain cases. By recent experiments one pound falling one foot, produced a pressure on a spring balance of nineteen pounds. Hence was discovered this general formula for the effect of a falling weight, viz.; $W\ (1 + 2\frac{1}{4} \times v) = W\ (1 + 18\ \sqrt{h})$. *e. g.* one thousand pounds falling nine feet produces a pressure $= 1{,}000\ (1 + 18\ \sqrt{9}) = 55{,}000$.

This result might vary with every different spring used. The impact or force of ice in motion is greater in appearance than in reality.

Pile Driving.

11. In pile driving, letting w = weight of ram, and w' = weight of pile, h = height of fall, v = velocity of ram when it strikes the pile. Then u = common velocity of ram and pile after fall $= \frac{w\,v}{w + w'}$. Their "*vis viva*" =

$$\frac{w + w'}{g} \times u^2 = \frac{w + w'}{g} \times \left(\frac{w\,v}{w + w'}\right)^2 = \frac{w^2\,v^2}{g\,(w + w')}$$

The "potential energy" of the ram and pile, or the work which this "*vis viva*" is capable of, equals half of it $= \frac{w^2}{w+w'} \times \frac{v^2}{2\,g} = \frac{w^2\,h}{w+w'}$. Putting this under the formula $\frac{w\,h}{1+\frac{w'}{w}}$ we infer; 1st. That the effect is directly proportional to the fall; 2d. That for different falls and weights, such that $w\,h$ remains the same, the effect will be greater the heavier the ram with a corresponding smaller fall. e. g., letting $w = 2$ tons, $w' = \frac{1}{4}$ ton, $h = 10$ feet, the effect, $= \frac{20}{1 + \frac{\frac{1}{4}}{2}} = 17.7777 +$. Let $w = 1$ ton, $w' = \frac{1}{4}$ ton, $h =$ 20 feet, effect $= \frac{20}{1 + \frac{1}{4}} = 16$.

For the resistance which the friction of ground opposes to the sinking of a pile under a weight, see Part IV.

Water in motion, or Waves.

12. The greatest force observed in the Atlantic, on the west coast of Scotland, was six thousand pounds per square foot. The average pressure was in summer six hundred pounds, and in winter two thousand pounds.

Air in motion, or Wind.

13. Experiments give the pressure in pounds, on a plane surface perpendicular to the wind, as, =, .0028 A. $v.^2$ = a little more than $\frac{1}{400}$ × A. $v.^2$ v being velocity of wind in feet, per second, and A the area of surface in square feet. If v = velocity in miles per hour then pressure = .0049 A. $v.^2$.

THE WIND.	Velocity in ft. per sec'd.	Velocity in mls per hour	Pressure in lbs. per sq. ft.
Fresh wind	20	13	0.95
High wind...............	40	26	4.00
Violent wind..............	60	40	8.00
Tempest..................	80	54	16.00
Violent tempest	100	68	24.00
Hurricane.................	120	80	32.00
Greatest hurricane	150	102	55.00

It is safe in temperate climates to take forty pounds per square foot as the greatest force to be guarded against. Some take thirty pounds.

The force operates horizontally, and its resultant passes through the centre of gravity of the surface. The pressure against a cylinder is from two-thirds to one-half the force corresponding to its diametrical section.

The effect of air in motion against a body : the resistance experienced by the body moving through the air with the same velocity . : : 1.4 : 1 .

PART II.

STRENGTH OF COHESION, OR THE RESISTANCE OF MATERIALS TO PARTING.

Preliminary Principles.

14. *Stress* is the force which acts upon a body. Its intensity is usually expressed as so many pounds per square inch. In French measures as kilogrammes per square millimetre, which equals 1.422 pounds per square inch.

The effect of a stress upon a body is its *Strain*, i. e., its alteration of volume and figure. When the strain is so great as to separate a solid body into parts, this constitutes its *fracture*.

Bodies resist stress by their elasticity, which is the property they possess of resisting any change in their form or volume. They may be regarded as composed of molecules which attract and repel each other with forces which in the normal state of the body are equal. Consequently every effort to separate these molecules calls into play an attractive reaction, and every effort to press them together calls forth a repulsive reaction which is proportional to the stress.

Within certain limits the elongations, or contractions, produced by any force, are nearly proportional to those forces; beyond those limits proportionality ceases, and the elongations and contractions increase more rapidly than the forces which produce them. This limiting point is called the limit of elasticity. Within the limits of proportionality, when the force ceases acting, the effect also ceases and the body returns to its original shape and dimensions, or nearly so. Beyond those limits it does not so return, but remains permanently elongated, contracted or bent.

It is then said to have its elasticity impaired or injured, or to have taken a "set." A long continued stress produces much more effect in impairing the elasticity of bodies than does a much greater force applied for a short time.

A body should never be subjected to an effort capable of seriously injuring its elasticity, for that is an approach toward fracture. The allowable strain should be very much less than this, on account of accidental forces and chemical changes to which a body is exposed. Usually the stress intentionally applied to a body is from one-third to one-quarter of that which would cause fracture, or one-half of that which corresponds to its limits of elasticity.

The following are the corresponding stress, strains and modes of fracture:

Stress.	Strain.	Fracture.
Pull	Extension	Tearing.
Shearing	Distortion	Shearing.
Thrust	Compression	Crushing.
Bending	Bending	Breaking across.
Twisting	Torsion	Wrenching.

A *pull* is a longitudinal stress which tends to lengthen a body.

A *thrust* is a longitudinal stress which tends to shorten a body. Usually a pull has a positive algebraical sign and a thrust a negative.

The other stresses are more or less transverse, and tend to distort the body.

For shearing stress see Art. 30.

For bending stress see Art. 36.

For twisting stress see Art. 84.

Resistance of bodies to Extension.

General Principles.

15. The resistance of bodies to extension is sometimes called their tensile, their direct, their absolute or their positive strength.

Let a prismatic body of a length l, and a cross section A, be subjected to a stress or pull P. It is elongated by a length δl. Let the ratio $\frac{\delta l}{l}$ equal i, and be constant within certain limits. Within those limits i is proportional to $\frac{P}{A,}$ i. e., to the stress per unit of cross section of the body; so that $\frac{P}{A}$ is to i, or $\frac{P}{A i}$ is a constant quantity. It is called the coefficient or modulus of elasticity, and it is expressed by E, so that $\frac{P}{A i} = E$. If we suppose the cross section o the body to be equal to the unit of surface, usually a square inch, and suppose it is possible for the elongation to become equal to the unit of length without injury to the elasticity, the body thus continuing to elongate in a constant ratio, we should have $A i = 1$, since $A i = A \times \frac{\delta l}{l} = 1 \times \frac{l}{l} = 1.$, and $P = E$ would then be the stress per unit of surface, capable of producing per unit of length, an elongation equal to that of the length. This may be taken as a definition for "*modulus of elasticity.*"

Thus if 20,000 pounds elongated a bar of wrought iron, whose section is the unit of surface, $\frac{1}{1500}$ of its length, then for this iron $E = 20{,}000 \times 1{,}500 = 30{,}000{,}000$ pounds. From $\frac{P}{Ai} = E$, we obtain where $A = 1, E = \frac{P}{i}$.

Experiments then on any material will give E for the unit of cross section. Knowing this we can calculate for a prismatic body of a given cross section A, the stress which will produce a given elongation, or the elongation which a given stress will produce, since the resistance of bodies to extension is directly proportional to their cross section, i. e., their section at right angles to the direction of the stress. Such experiments are best represented graphically, by constructing a curve of which the ordinate are the forces applied, and the elongations the abscissas. Figure 2 refers to a bar of iron elongated by successive weights. From the inclinations of the curve at various successive points we can infer the laws of the action of the strain; whether it be uniform or varied, increasing or decreasing, &c. The total resistance of a body to elongation is equal to the work done in producing that elongation, and equal to the sum of the products of the stresses exerted, multiplied by the elongations produced by them.

Fig. 2.

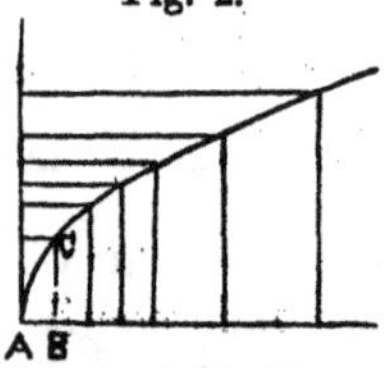

The area $\frac{1}{2}$ B C $\times$ A B, of the triangle A B C, is formed by the abscissa and ordinate, and the first straight line portion of the curve; therefore expresses the work done by the stress applied. The mean stress up to that point of strain $= \frac{1}{2}$ B C × A B ÷ A B $= \frac{1}{2}$ B C $= \frac{1}{2}$ final stress.

So to, the quadrature or area of the curve, up to any point, gives the work done in producing the elongation corresponding to that point. Thus the quadrature up to the so called limit of elasticity, gives the work developed up to that point, by the elasticity of the body.

It is called *The Living Resistance of Elasticity*, and it is written T_e.

The quadrature, up to the point of rupture, gives the work done in producing rupture. It is called *The Living Resistance of Rupture*, and is written T_r.

To resist with safety a very sudden pull a bar or rod requires twice the strength that is necessary to resist the gradual application and steady action of the same pull.

For the work done by a constant force $\frac{1}{2}$ P, acting through a given space, equals the work done by the action through the same space of a force increasing uniformly from 0 to P.

Special Materials.

Wrought Iron.

16. The following table shows the relative weights and elongations of a rod of good wrought iron:

Load per square inch in lbs.	Elongation per unit of length.	Permanent elongation per unit of length.
5,300	.00018	.000000
12,300	.00047	.000004
21,300	.00076	.000010
26,700	.00101	.000083
29,900	.00128	.000263
32,000	.00229	.001130
34,000	.00429	.003071
40,000	.01049	.009102
50,700	.03493	.03280
52,300	Rupture.	Rupture.

Representing this table by a curve, as in Fig. 2, we infer:

1st. Since the first portion of A is a straight line, the ratio of elongation to the load is so far constant. It continues so up to a stress of about 17,000 pounds per square inch.

2d. For a stress greater than this the line tends towards the axis of abscissas. This shows that the elongations are increasing faster than the loads.

The point where this change takes place is called "the limit of elasticity."

3d. Beyond the stress of about 30,000 pounds, we have nearly a straight line again, but much more inclined toward the axis of abscissas. This shows that the elongations have again become proportional to the loads, but have a much greater ratio to them than at first, showing that the elasticity of the material is injured. The permanent elongations follow similar laws, they are very small up to a pull of 21,000 pounds, being then only .00001, but then increase very rapidly, much more so than the temporary elongations.

In comparing different irons we find that the weights which alter the elasticity of soft bars do not do so for soft wire; also, that this wire breaks with a greater load but less elongation; that hard wire breaks with somewhat more weight than the soft, but with much less elongation, and that it gives way suddenly and without warning; we also find that the hard iron has much more T_e than the soft iron, i. e., that greater stresses may be applied before the limit of elasticity is reached; but soft iron has much more T_r than the hard iron, i. e., much more work must be done to break them; therefore they should be chosen for exposure to sudden shocks.

Example. In soft iron, if the limit of elasticity be 22,800 pounds, and the elongation be .00086, we have $T_e = \frac{1}{2} \times 22{,}800 \times .00086 = 9.8$ pounds per foot of length.

Suppose a bar be twenty feet long and one inch square, its $T_e = 20 \times 9.8 = 196$ foot pounds. Now if a body W falls from a height h the work of that body $= \frac{1}{2} m v^2 = \frac{1}{2} \frac{W}{g} \times v^2 = Wh$. Suppose the body to be suspended from the bar, this work is to be destroyed by the Living Elasticity T_e, which is here one hundred and ninety-six pounds, which amount must not be exceeded. It would be equal to the work of a body of ten pounds falling 19.6 feet. Beyond that the elasticity would be altered. The total quadrature $T_r = 6{,}400$. Then the above bar would be broken by one hundred pounds falling from a height equal to $\frac{1}{2} \times \frac{6400}{100} \times 20 \times 1 = 640$ feet.

In like manner soft annealed wire gives $T_e = 9.4$ and $T_r = 710$. For hard unannealed wire $T_e = 8.3$ and $T_r = 970$. These are for one inch square. Very extensible materials resist shocks much better than rigid ones. The strength of wrought iron varies greatly with its chemical composition, mode of manufacturing, size of bar, &c.

Cast iron; stress per sq.. in., in lbs.	Wrought iron; stress required to produce the same elongat'n	ELONGATION.		"Set" cast iron	"Set" wrt. iron	Ratio
		Cast iron	Wrt. iron			
512	1,254	.00004	same	0	0	......
2,486	5,600	.0002	same	.00001	0	......
4,480	10,080	.00033	same	.000026	.0000049	1:15
5,600	12,544	.00042	same	.000036	.000006	1:17
6,720	15,142	.00052	same	.000047	.0000075	1:16
8,960	20,160	.00072	same	.000075	.0000225	1:30
12,330	26,970	.00107	same	.000124	.0001166	1:94
13,216	29,680	.00121	same	.000160	.0003583	1:222
......	33,600	......	.0023	......	.001	.....

Relative tensile forces to produce equal elongations in wrought and cast iron is two and one-quarter to one. Within the above limits, for equal elongations, the "*set*" of wrought iron is less up to strains of 12,000 pounds for cast and 27,000 for wrought iron, which produce equal elongations. Beyond that the "*set*" for wrought iron is much the greater.

For a stress of about two-thirds rupturing stress of each, (e. g. a stress of five tons for cast and fifteen tons for wrought iron,) the elongation for the wrought iron is two and a-half times that of cast iron, and its "*set*" ten times that of cast iron, e. g.:

⅓ rupturing wt. = stress.	ELONGATIONS.			SETS.		Ratio.	Ratio of set to elongation.
	Cast.	Wro't.	Ratio.	Cast.	Wro't.		
11,120	.00095	...	1:24	.00011	......	1:10	1:9
33,600	......	.0023	....	.....	.0011	....	1:2

Wrought iron is also affected by temperature. Its strength increases up to 300° Fah., and then decreases, being only about half as strong at 1000° as at 300°.

Cast Iron.

17. The resistance of cast iron to breaking by extension is much less than that of wrought iron, but follows similar laws.

Other Metals.

18. For "Table of Resistance to Extension" see following table :

TABLE OF RESISTANCE TO EXTENSION.

METAL.	Stress in lbs. per sq. in. corresponding to limit of elasticity.	Elongations at limit of elastic'y.	E = coefficient or modulus of elasticity.	R = stress in lbs. per sq inch causing rupture.	
				Extreme.	Safe means for sound material.
Iron.					
Bars	17,000	1:500 1:200	29,000,000	35,000 85,000	48,000
Boiler plate..	26,000	......	....	46,000 56,000	51,000
Annealed wire	20,000	1:1250	25,000,000	*70,000 130,000	....
Steel	35,000 94,000	1:830 1:450	29,000,000 42,000,000	70,000 130,000	90,000
Copper.					
Cast.........	....	......	....	....	19,000
Bolt	....	......	....	....	36,000
Sheet........	....	......	....	....	30,000
Wire	....	......	16,000,000	....	60,000
Brass.					
Cast.........	7,000	1:1300	9,000,000	....	18,000
Wire.........	19,000	1:740	14,000,000	....	49,000
Bronze or gun metal	6,000	1:1600	10,000,000	....	36,000
Tin (cast) ...	....	......	4,500,000	....	4,600
Zinc	....	......	13,500,000	....	7,500
Lead.					
Cast.........	1,400	1:500	700,000	....	1,400
Wire	570	1:1800	1,000,000	....	....
Sheet........	....	......	700,000	....	3,300
Cast iron	8,500 14,000	......	14,000,000 23,000,000	13,000 29,000	16,500
Corning's semi-steel...	....	......	....	73,000 89,500	....

* Small wires are much stronger per square inch than larger ones. For wire 20 feet to a pound, strength = 1,300 pounds = 90,000 pounds per square inch. Wire one-eighth in diameter breaks at 1,600 pounds = 130,000 pounds per square inch.

The safe stress for extension of metals is from one-sixth to one-third of that causing rupture, depending upon their exposure to shocks, weather, &c. Some recommend taking half the stress corresponding to the limit of elasticity. These two methods do not vary much in their results. When they do adopt the safer one.

Pure cobalt is the most ductible and tenacious of metals, a wire of it is twice as strong as an iron wire. Copper (rolled) is weakened by heat above the freezing point. Aluminum has tenacity of 18,600 pounds, and hence is between zinc and copper.

An alloy of ninety parts copper and ten aluminum has a tenacity of 80,000 pounds, hence between soft iron and steel.

Captain Meigs, at the capitol, allowed for greatest safe strain for iron as follows: For important parts, cast iron 1,800, and wrought iron 8,000. For secondary parts, cast iron 3,000, and wrought 9,500. From 8,000 to 10,000 is usually allowed for wrought iron in bridges.

Length of bars which would break of their own weight: Steel 40,000′; wrought iron 18,000′; cast iron 5,000′; rolled copper 9,000′; lead 360′; cast copper 5,000′.

Resistance of Wood to Extension.

19. This, as for metals, is proportional to the cross-section of the pieces. In wood even more than in metals there is strictly no limit of elasticity, but any load produces a permanent elongation, but in dry woods very much less than in wet woods.

KIND OF WOOD.	Stress corresponding to limit of elasticity.	Elongation at limit of elasticity.	E = modulus or coefficient of elasticity.	R = stress causing rupture.
Ash	1,800	1:855	1,600,000	17,000
Beech	2,300	1:570	1,300,000	11,000
Birch	....	....	1,600,000	15,000
Chestnut......	....	....	1,100,000	11,000
Elm	3,300	1:414	1,360,000	14,000
Larch	2,500	1:520	900,000 } 1,300,000 }	10,000
Locust	....	....	....	16,000
Oak	2,850	1:600	1,700,000	10,000 } 15,000 }
Red pine	4,500	1:470	2,100,000	13,000
Yellow & White pine.........	3,100	1:850	2,610,000	10,000

One-fifth to one-tenth of *R* is considered a safe load, depending on importance, exposure, &c. When exposed to running loads, as in bridges, one-twelfth is better; say for white pine, eight hundred pounds per square inch. The above numbers are for sound timber, free from knots and other defects. The strength of the same species of wood varies greatly according to the locality of the tree, nature of the soil, part of the tree, age of tree, its seasoning, &c.

The strength of wood, as given from the above table, supposes it to be pulled in the direction of its fibres or with the grain. At right angles to this its strength is much less. In straight grained woods like pine, this last strength is only from one-sixteenth to one-twentieth of the former. In tough and crooked woods it is from one-seventh to one-tenth. Care must be taken when calculating the strength of wood to resist extension, to allow for the parts which must be cut away by notches, bolt-holes, &c., in order to make the connections.

Other Materials.

20. Stone is rarely exposed to a strain of extension, its resistance thereto ranges from two hundred to one thousand pounds per square inch. The safe strain is one-tenth of that.

Brick one hundred to three hundred pounds. Glass two thousand four hundred to two thousand nine hundred pounds. The poorest mortar resists with only ten pounds per square inch. Good hydraulic mortar with from one hundred to two hundred pounds per square inch. Safe strain one-tenth in all cases. Adhesion to stone of good mortar, fifteen to thirty-three pounds. Plaster, thirty to forty-eight pounds. Best Rosendale cement, forty pounds.

TABLE (RANKINE)

MATERIAL.	"E"	"R"
Bricks.................	....	280 to 300
Glass.......	8,000,000	9,400
Slate..................	13,000,000	10,000 to 13,000
Ordinary mortar........	....	50

PARTICULAR FORMS.

Riveted Iron Plates.

21. The line of fracture of a riveted plate, is a line drawn on it crossing all the lines of strain in such a manner that the section of the plate along it shall be a mini-

mum. The plate must break along this line, however "zig-zag" it may be.

The strength of the rivets will be considered under "Shearing stress."

Wire Ropes.

22. Their breaking weight averages four thousand five hundred pounds per pound weight per fathom, and the working load one-sixth of this.

TABLE OF WIRE ROPE MANUFACTURED BY JNO A. ROEBLING, TRENTON, N.J.

ROPE OF 133 WIRES.						ROPE OF 49 WIRES.				
Trade number.	Circumference in inches	Diameter.	Price per foot, in cts.	Ultimate strength in tons of 2,000 pounds.	Cicumference of Hemp Rope of equivalent strength in inches.	Trade number.	Circumference in inches	Price per foot, in cts.	Ultimate strength in tons of 2,000 pounds.	Circumference of Hemp Rope of equivalent strength in inches.
1	6¾	2¼	1 20	74 00	15½	11	4⅝	54	36 00	10¾
2	6	2	1 05	65 00	14½	12	4¼	47	30 00	10
3	5½	1¾	91	54 00	13	13	3¾	41	25 00	9½
4	5	1⅝	78	43 60	12	14	3⅜	35	20 00	8¼
5	4¾	1½	65	35 00	10¾	15	3	29	16 00	7¼
6	4	1¼	53	27 20	9½	16	2⅝	23	12 30	6¼
7	3½	1⅛	41	20 20	8	17	2⅜	18	8 80	5½
8	3⅛	1	34	16 00	7	18	2⅛	15	7 60	5
9	2¾	⅞	28	11 40	6	19	1⅞	13	5 80	4¾
10	2¼	¾	25	8 64	5	20	1⅝	11	4 09	4
10¼	2	⅝	24	5 13	4½	21	1⅜	9	2 83	3¼
10½	1⅝	9-16	23	4 27	4	22	1¼	8	2 13	2¾
10¾	1½	½	22	3 48	3¾	23	1⅛	7	1 65	2½
For safe working load allow 1-5 to 1-7 of ultimate strength.						24	1	6½	1 38	2¼
						25	⅞	6	1 03	2
						26	¾	5½	0 81	1¾
						27	⅝	5	0 56	1½
						27½	½	4		

Hempen Ropes.

23. Breaking weight 5,500 to 12,000 pounds per square inch of cross-section. Average of good rope 8,500 pounds. Safe strain one-half the breaking weight.

Another Rule.

Safe strain in pounds equals circumference in inches squared, × 200 = cross-section in inches × 2500.

Another gives: Breaking weight equals one gross ton per pound weight per fathom.

Chain Cables.

24. When the links are formed as in figure 3, their strength is nearly equal to the strength of the bars of which the links are formed, i. e., equal to twice the section of the single bar. Some U. S. experiments gave a mean value of $R = 41,000$ pounds; extremes = 32,000 and 50,000.

Fig. 3.

Thin Hollow Cylinders.

25. Let p = the interior pressure per square inch, d = diameter in inches, R = resistance to tearing per square inch, and t = thickness in inches. Then the equation for equilibrium is, $pd = 2\,t\,R$. Hence, $t = \frac{p\,d}{2\,R}$.

For safe thickness we take $t' = \frac{p\,d}{2\,R'}$; in which R' means safe strain per square inch.

The pressure is often named as so many "atmospheres," each atmosphere being fifteen pounds per square inch.

Let n = number of atmospheres of pressure above that of the air, then the above formula becomes $t' = \frac{15\ n\ d}{2\ R'}$.

For cast iron, taking $R = 16{,}500$ pounds, the proof strain for water and gas pipes may be one-third of this, and the working strain one-sixth of it. In addition to the thickness necessary to resist internal pressure a constant thickness is added to resist external shocks. Calling this additional thickness t'' the formula for safe thickness becomes $t' = \frac{p\ d}{2\ R'} + t'' = \frac{15\ n\ d}{2\ R'} + t''$.

The Paris rule is, for cast iron take $R' = 3{,}100$ pounds and $t'' = 0.333$ inches.

For wrought iron, Paris rule makes $R' = 8{,}500$ pounds and $t'' = 0.12$ inches, or $t' = 0.00086\ n\ d + 0.12$.

A simpler formula for cast iron is this; $t' = 0.002\ n\ d + .4$; (all in inches).

For water pipes of rolled copper, $t' = .00147\ n\ d + .16$, (all in inches).

For water pipes of lead, $t' = .00242\ n\ d + .20$, (all in inches).

For water pipes of zinc, $t' = .0062\ n\ d + .16$, (all in inches).

For water pipes of wood, $t' = .0323\ n\ d + 1.08$, (all in inches).

For riveted plate iron (steam boilers) in the first general formula, $R = 34{,}000$ pounds.

The proof tension may be one-half of this, and the working tension one-eighth. For steam pipes the same; i. e., $R' = 4{,}250$ and $t' = \frac{p\ d}{2\ R'} + .12$.

The French Government rule is, $t' = .0018\ n\ d + .12''$. This about corresponds to the above rule.

The bursting strain on the longitudinal seams of cylindric boilers, is double the strain on the circular seams. Tubes are also liable to give way by the pressure from without, which makes them collapse. This is partly extension, partly crushing, and partly bending. It is usual to make the thickness of tubes exposed to collapsing, double of what would be required if only exposed to pressure from within.

Recent experiments show that the strength of such tubes varies directly as the square of the thickness, inversely as the diameter, and inversely as the length.

Mr. FAIRBAIRN'S rule, for the safe pressure in pounds per square inch, on flues of wrought iron is: multiply the constant quantity 800,000 pounds, by the square of the thickness in inches, and divide by the length in feet, and by the diameter in inches.

Thick Hollow Cylinders.

26. In these the full strength of the material is not obtained, because the inner ring may give way before scarcely any strain comes on the outer part. This is important in the case of hydraulic presses, cannons, &c.

Prof. BARLOW'S formula is $t = \frac{p\ d}{2 \times (R - p)}$, for bare equilibrium. Only one-third of the bursting pressure should be used in presses.

Spheres.

27. The strength of a sphere to resist bursting is pre-

cisely twice that of a cylinder of the same diameter. For cylinders we have $p\,d = 2\,t\,R, \therefore p = \frac{2\,t\,R}{d}$. For spheres $p = \frac{4\,t\,R}{d}$.

A Suspension Rod of Uniform Strength.

28. Let W = weight, suspended at lower end; let w = weight of rod per unit of volume; R' safe strain; x = length of rod up to any desired point; A = proper area or cross-section at the desired point. All the measurements must be in the same unit (say, inches). Formula

$$A = \frac{W}{R'}\,e^{\frac{w\,x}{R'}}$$

Here e is raised to a power whose numeration is w multiplied by x, and whose denomenator is R', $e = 2.71828$, (the Naperian base).

RESISTANCE TO SHEARING OR DETRUSION.

29. Shearing, or tangential stress, is the force which acts between two parts of a body, when each draws the other sideways, in a direction parallel to their surface of contact. It is the stress which effects rivets, trenails, notches at the ends of tie-beams, &c.

A shearing stress also accompanies bending and torsion to be examined hereafter. Resistance to shearing per square inch of cast iron is 24,000 to 42,000 pounds. Wrought iron 50,000 pounds. Safe, one-sixth of each. Chains for suspension bridges consist of long bars connected by pairs of short bars or links with bolts or pins passing through both.

The best proportions for economy should be such that the bars and bolts would be equally strained, so that neither should give way before the other. This is a fundamental principle in all combinations or parts of a structure; as in a chain all the links should be equally strong, since the strength of the chain is only that of the weakest link. To break the bolt it must be sheared across at two places at once, then the double cross-section of the bolt of diameter $d = 2 \times \left(\frac{11\,d^2}{14}\right) \times$ resistance to shearing per square inch, = cross-section of the bar, $a \times$ its resistance to tearing asunder. Calling these resistances equal we have $2 \times \frac{11\,d^2}{14} = a$. But these results are as 5:6. Then, $2 \times \frac{11\,d^2}{14}$: cross-section bar :: 6:5 $\therefore d = \sqrt{\frac{a}{1.31}}$ $= 874 . \sqrt{a}$.

30. Rivet work for bridges, plate boilers, &c., should also be so proportioned that the section of the uncut part, multiplied by its resistance to tearing per unit of measure, should equal the section of all the rivets, multiplied by their resistance to shearing.

Assuming that the rivets resist shearing with about the same strength that wrought iron plates resist tearing, the sections of the rivets should about equal the sections of the iron between them.

In the overlap plate joint, single-riveted, the sectional area of one rivet should equal the sectional area of the plate between the holes.

The diameter of the rivet is usually one and a-half to

two times the thickness of the plate. Calling t = thickness of the plate, d = diameter of the rivet, l = the distance between the holes; we have $l \times t = \frac{11\,d^2}{14}$. The distance between the rivets equals $\frac{11\,d^2}{14} \div t$, and the distance between their centres $= \left(\frac{11\,d^2}{14} \div t\right) d$. The overlap is about equal to the distance between their centres. In the overlap plate joint, double riveted, the sectional areas of two rivets equals the sectional area of the plate between each pair of holes in the same line.

The distance of centres $= \frac{11\,d^2}{7} \div t + d$.

The overlap is about 1.7 of this.

In a plate, butt joint, with two covering plates, single-riveted, the section of the rivets which give way by being sheared across in two places at once, is as in the preceding case.

Distance of centres the same $= \frac{11\,d^2}{7} \div t + d$.

Length of each covering plate about twice this.

In a plate, butt joint, double-riveted, section of four rivets must equal that of the plate between two holes in the same line. Distance of centres $= \frac{22\,d^2}{7} \div t + d$. Length of covering plate equals three and one-third to three and one-half times this.

(On rivet work see Latham on wrought iron bridges, pp. 26-36.)

Wood.

31. When a rafter is notched into a tie-beam, the end of the beam is liable to give way by shearing off or detrusion.

Fig. 4.

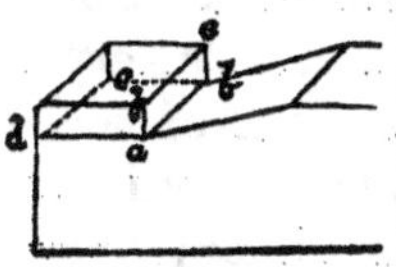

For greatest economy of material the resistance to thrust or compression on the surface *a b e f*, Fig. 4, should just equal the resistance to shearing of *a b c d*. Calling the former resistance ten times the latter, *a d* should be ten times *a f*. If the shoulder should be in the middle of the beam, as for a bolt-hole, there would be twice the resistance to detrusion.

Resistance to detrusion of Wood with the fibres.

	lbs. per sq. in.
White pine	490
Ohio pine	390
Georgia pine	410
Spruce	470
Hemlock	540
Chestnut	690
Oak	780
Locust	1,180

Across the fibres.

Pine	400 to 800
Spruce	600
Larch	1,00 to 1,700
British oak	2,300

Safe stress, one-quarter to one-sixth of the above.

Resistance to Compression and Crushing.

32. For small pressures, this resistance equals the resistance of the same material to extension, and has the same numerical modulus of elasticity, which here means the weight, which continuing to act in the same ratio would shorten the body to one-half of its length. For greater pressures the resistance to compression is very irregular. It varies directly as the cross-section of the material. Different materials yield in different ways. Granular substances, like cast iron, stone, brick, &c., give way by oblique shearing, at a certain angle with the direction of the crushing force, which angle in cast iron is from 30° to 40°. Sometimes a wedge-shaped piece or pyramid is forced out on two or four sides. Substances of a glassy texture give way by splitting irregularly. Tough and ductile substances, like wrought iron, yield by bulging or swelling out sideways. Fibrous substances, like wood, yield by buckling the fibres, wrinkling and splitting.

When the material compressed is a certain number of times longer than it is thick, it gives way by bending or cross-breaking. The resistance varies inversely with the length (nearly). This case will be examined under Part IV.

When the centre of pressure upon a surface does not coincide with its centre of figure, the maximum pressure will exceed the mean pressure. Considering the pressure to be a uniformly varying stress, the ratio of maximum to minimum pressure equals $1 + b\,c$ to 1, in which $b =$ distance of centre of pressure from centre of figure, and

c = a quantity depending on the form of a section of the body at right angles to the direction of the force.

For a square of side h or rectangle of thickness h, in a direction perpendicular to the central line, $c = \frac{6}{h}$.

For a circle of diameter h, $c = \frac{8}{h}$.

When the pressure is at one-third h from the edge, or one-sixth h from the centre, then the ratio is $\left(1 + \frac{1}{6} h \times \frac{6}{h}\right) : 1 :: 2:1$, i. e., the maximum pressure equals twice the mean pressure. If the pressure be at $\frac{1}{6}$ h from the edge, or $\frac{1}{3}$ h from the centre, the ratio equals $\left(1 + \frac{1}{3} h \times \frac{6}{h}\right) : 1 :: 3:1$. The pressure = 0 at $\frac{1}{3}$ h from the other side. If the pressure be at the edge, or $\frac{1}{2}$ h from centre, then ratio is $\left(1 + \frac{1}{2} h \times \frac{6}{h}\right) : 1 :: 4:1$.

Stone. (*Prof, Henry.*)

Name of Stone.	Resistance to crushing in lbs. per sq. in.
Sandstone (U. S. capitol)	5,200
Red sandstone (Smithsonian Institute)	9,500
Marble	7,000 to 10,000
Malone sandstone	24,000
Blue gneiss	15,000
Quincy granite or sienite	29,000
Brown sandstone or freestone	3,000 to 3,500
New Jersey freestone	3,500
Connecticut freestone	3,300
Dorchester freestone	3,000
Albert sandstone	8,200
Caen stone	1,100

Britania Bridge Experiments.

Anglesea limestone, ... 7,600
Red sandstone, ... 2,180
Brick masonry, 417 to 610 mean = 521

New brick walls of Capitol 1,333 pounds per square inch.

White marble of Washington Monument 2,000 to 5,000 pounds per square inch.

British Parliament Experiments.

STONES.	Fracturing weight, in lbs. per sq. in.		Crushing weight, in lbs. per sq. in.	
Sandstone	1,792 6,049	mean 3,584	3,584 7,840	mean 5,824
Magnesium limest'e	4,928 6,720		3,808 8,288	mean 5,152
Limestones	1,120 1,568	mean 1,344	1,792 4,032	mean 2,912
Silicious limestones.	2,912		7,168	
Oolite	1,344 2,016	mean 1,568	1,568 3,808	mean 2,688

Bramah's Experiments.

STONES.	Fracturing weight.		Crushing weight.	
Granites...........	6,496 10,752	mean 8,288	9,856 14,784	mean 11,872
Sandstones	2,240 6,496	mean 2,698	2,464 8,736	mean 3,800

Good brick work in cement, cracked at 760, crushed at 900. Common brick crushes at from 900 to 1,900 pounds per square inch. Soft brick with from 460 to 620.

The strength of stone remains constant till its height equals ten or eleven times its diameter. When height is twenty-four times diameter its strength equals seven-tenths of original strength. At thirty times, it is $\frac{44}{100}$, and at forty times, $\frac{37}{100}$. The resistance of stone to compression is about ten times that to extension.

A granite column would be crushed by its own weight when about 12,000 feet high; white marble 2,000.

In the greatest buildings in the world, the greatest pressure is from one-sixteenth to one-eighth the crushing weight; one-tenth is a good ratio.

Common mortar crushes with from 300 to 500 pounds per square inch. Common hydraulic mortar, 1,000 pounds; Very hydraulic mortar 1,500 pounds.

Concrete made from quick lime will bear safely 150 pounds per square inch. This was the foundation of the London Crystal Palace, where it bore forty pounds. 'Beton' (concrete made of hydraulic cement) crushes with 600 pounds per square inch. Safe pressure, one-tenth of this. Masonry should not be loaded with more than one-twentieth the weight, which would crush the material, or one-tenth of what would crush the masonry.

Cast Iron.

33. The resistance of cast iron to crushing is from 80,000 to 140,000 pounds per square inch, the mean being 110,000. This is about six times its resistance to extension. One-fifth the crushing weight is safe.

The English cast iron bridge engineers use 18,000 pounds. Captain MEIGS makes the limit for pieces less than twelve times as long as thick, for principle parts,

10,500 pounds per square inch, and for secondary 17,500. Where the length is more than three times the thickness the pieces give way by bending. It is brittle at the freezing point. Its strength varies little between 40° and 120°, beyond 120° it becomes weaker.

Wrought Iron.

34. Its crushing weight is from 35,000 to 40,000 pounds; one-half to one-third of that of cast iron. This is about three-fourth its resistance to extension. The safe load is one-fourth of this. Captain Meigs, for pieces less than twelve times as long as thick, uses 7,000 pounds.

Great care must be taken to prevent the bending of long pieces, subjected to thrust, by supporting them at different points in their length, or by putting it in the form of tubes, &c., or by corrugating thin plates of it, or by riveting T irons to the plates. Although wrought iron is crushed with much less weight than cast iron, yet under moderate pressure (within the limit of elasticity,) it is compressed much less than cast iron. Its modulus of elasticity being almost double. Therefore it should be preferred in important structures, even for resisting compression, except where economy forbids. It was used in the Britannia tubular bridge.

Crushing weight of cast brass equals 10,300 pounds per square inch.

Wood.

35. Dry timber crushed along the grain yields under a pressure of from 5,000 to 10,000 pounds per square inch. Dry English oak, elm, beech, ash, 9,000 to 10,000. Pine

5,000 to 7,000 pounds; one-tenth of crushing weight is safe. When green its strength is very much less, sometimes one-half. For good white pine 800 to 1,000 pounds per square inch may be used for short pieces.

Sheet lead or iron shoes should be interposed to prevent the fibres working into one another, where the end of one piece of wood presses against another. Where a pressure acts transversely to the fibres, if it acts on the whole side of the piece, 100 to 250 pounds is the safe limit, or say one-fourth or one-fifth of the former. When a wooden post or strut is much longer than it is thick, it yields by bending. Its strength decreases rapidly as its length increases. (See notes on bending.)

Resistance to Bending and Breaking.

General Principles of Flexure.

36. Let AB be a body, such as a beam, fixed at one end A, and having at the other end B, a weight, or some other force, applied perpendicularly to its length. The weight of the body itself is neglected for the present. This force exerts a shearing stress equal to W, at every point of the beam, acting at right angles to it, and also a bending force along the beam. The beam will therefore assume a curved form, as shown by the dotted lines in the figure; B coming to B′. The upper fibres will be lengthened, and the lower ones shortened, while one layer of fibres will be neither lengthened nor shortened. This line is therefore called the "*Neutral Axis*," because along it the bending forces are neutralized, and the shearing force alone is acting. This flexure calls out a reaction of the fibres, and it will continue till the elastic forces of the body, by their reaction, come into equilibrium with the weight or other force. The fibres of a beam are lengthened or shortened in proportion to their distance from the neutral axis. Let v_1 = this distance for any extended fibre, and r = radius of curvature for that part of the body.

Fig. 5.

Fig. 6.

Fig. 6 is an enlarged section of a portion of a beam. AB is the neutral axis. Then by similar triangles we have $l : r :: \Delta l : v_1$ and the elongation per unit of length $\frac{\Delta l}{l}$, expressed by i, $= \frac{v_1}{r}$.

The force P necessary to produce the elongation $= Ei = E \times \frac{v_1}{r}$, and for an elementary section $d\,a$ of a body of section a, it is $E\frac{v_1}{r}\,d\,a$. This expression is the "*elastic reaction.*"

For any compressed fibre we have in like manner, calling v_2 its distance from the neutral axis, its elastic reaction equals $E\frac{v_2}{r}\,d\,a$.

The first condition of equilibrium is, that the sum of the reactions of all the extended fibres should equal that of all the compressed fibres, i. e.,

$$\int\frac{E\,v_1\,d\,a}{r} = \int\frac{E\,v_2\,d\,a}{r} \quad [1].$$

From the above we get $\int v_1\,d\,a = \int v_2\,d\,a$, which shows that the neutral axis passes through the centre of gravity of the cross-section of the body.

The "*moment*" of each fiber is the product of its elastic reaction, by its leverage, i. e., it is $= \frac{E\,v_1\,d\,a}{r} \times v_1$, or $\frac{E\,v_2\,d\,a}{r} \times v_2$. Then the moment of all the fibres equals the integral of the above $= \frac{E}{r}\int v_1^2\,d\,a + \frac{E}{r}\int v_2^2\,d\,a$, or calling the distance from the neutral axis v, positive or negative, according as the fibres are compressed or extended, then the expression becomes $\int\frac{E\,v\,d\,a}{r} = 0$.

This last expression is called the "*moment of elasticity.*" To find an expression for r, the radius of curvature, the equation of a curve being $y = \int(x)$, its radius of curvature

$$r = \left[1 + \left(\frac{d\,x}{d\,y}\right)^2\right]^{\frac{3}{2}} \div \frac{d^2\,y}{d\,x^2}.$$

In the case under consideration, the flexure or bending is very slight, so that $\frac{d\,y}{d\,x}$ (which is the tangent of the slope) may be neglected when compared with unity, and we get $r = \frac{d\,x^2}{d^2\,y}$; and the moment of elasticity becomes

$$E\frac{d^2\,y}{d\,x^2} \times \int v_1^2\;d\,a + E\frac{d^2\,y}{d\,x^2} \times \int v_2^2\;d\,a. \quad [2.]$$

When the piece is prismatic or cylindrical throughout its length, the above integral is constant for every section, and is its moment of inertia. Call this I, and the above moment of elasticity becomes $E\,I\frac{d^2\,y}{d\,x^2}$.

The second condition of equilibrium is, that the moment of elasticity of the fibres shall equal the moment of the external forces which tend to bend the piece. This latter moment will be designated by M.

FLEXURE OF PRISMATIC BEAMS.

When fixed at one end.

37. When fixed at one end and loaded at the other, as in Fig. 7. $A\ B$ represents the middle fibre of the beam, fixed at A, and having a force at the other end, B, applied perpendicularly to its length. Its length is l, and is supposed to be sensibly the same before and after being bent. The shearing force is constant for all points of the beam and equals W. The bending moment of the force at any point, distant from the fixed end x units, or x' from the other, $= W(l - x), = W(x')$.

Fig. 7.

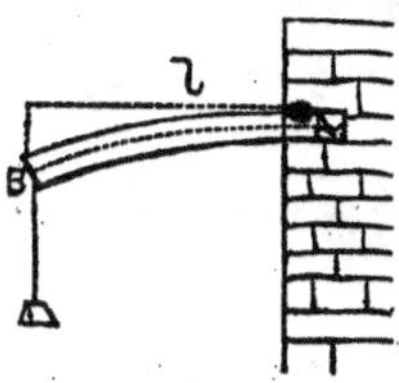

To find the deflection y at any point distant x from the origin: equate the moment of elasticity of the beam, and the moment of the bending forces; i. e., $EI\frac{d^2y}{dx^2} = W(l - x)$. By integrating twice we get $EI(y) = W\left(\frac{lx^2}{2} - \frac{x^3}{6}\right)$. Whence $y = \frac{W}{EI}\left(\frac{lx^2}{2} - \frac{x^3}{6}\right)$.

At the end of the beam y becomes δ and x becomes l, and we have $\delta = \frac{Wl^3}{3EI}$. [3.]

NOTE.—The value of E directly obtained by experiment on extension, and as given in the tables, is smaller than its proper value, as found from direct experiments on bending, owing to the increased resistance which the cohesion of adjacent layers of fibres oppose to sliding on one another. Hence the former value of E must be modified accordingly before being used in this and the following formulas.

When the beam is fixed at one end and loaded uniformly along its length. Fig. 8.

Fig. 8.

Let $w =$ load per unit of length, so that $w\,l =$ whole load W. The moment of this load at any point, x from the origin is $w\,(l - x)\,\frac{1}{2}\,(l - x) = \frac{1}{2}\,w\,(l - x)^2 = \frac{1}{2}\,w\,(x')^2$.

Equating and integrating as in the preceeding case we get

$$y = w\frac{(\frac{1}{4}\,l^2 x^2 - \frac{1}{6}\,l\,x^3 + \frac{1}{24}\,x^4)}{E\,I} \text{ and } \delta = \frac{w\,l \times l^3}{8\,E\,I} = \frac{w\,l^4}{8\,EI} \quad [4]$$

Comparing this expression for δ with equation [3], we see that the deflection is three-eighths what it was in the former case.

When there is a uniform load acting downward as before, and also an upward force, P, at the end. Let $x_1 = l - x$. The bending moment $M = w\,x_1\,(\frac{1}{2}\,x_1) - P\,x_1 = \frac{1}{2}\,w\,x_1^2 - P\,x_1 = \frac{1}{2}\,w\,x_1\,(x_1 - \frac{2\,P}{w})$

When x_1 is less than $\frac{2P}{w}$, M is negative, and the curve of the beam is concave upwards.

When x_1 is greater than $\frac{2\,P}{w}$, M is positive, and the curve has an inflexion at the point where $x_1 = \frac{2P}{w}$.

A beam supported at both ends. Fig. 9.

Figs. 9 and 10.

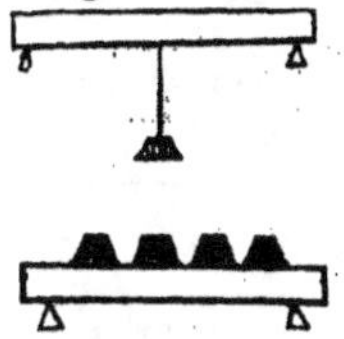

38. Let it be loaded in the middle with W. The pressure on A and B is half W, and the reaction of each is the same; each half of the beam is therefore in the same condition as a beam fixed at one end and loaded at the other with half W. Then for any point, x from the centre, the moment of the bending force $= \frac{W}{2} (\frac{1}{2} l - x)$.

Equating as before and integrating from 0 to $\frac{1}{2} l$ we get

$$\delta = \frac{W l^3}{48 E I}. \quad [5.]$$

That is, the deflection is one-sixteenth of that of a beam of the same length, and having the same weight applied at its extremity. The same beam with a uniform load, Fig. 10, would give, putting $w l$ in place of W;

$$\delta = \tfrac{5}{8} \times \frac{w l \, (l^3)}{48 E I}. \quad [6.]$$

That is, five-eighths of what it was with an equal load at the middle. This beam, compared with a beam fixed at one end and loaded uniformly, gives the ratio $\frac{3}{8} : \frac{5}{8} \times \frac{1}{16}$ or $1 : \frac{5}{3} \times \frac{1}{16} = \frac{5}{48}$.

39. *A beam fixed at one end and supported at the other.*

Fig. 11.

With a load W in the middle, Fig. 11, the deflection at the lowest point is, $\frac{45}{100}$ of the deflection of the same beam merely supported at both ends. The lowest point, or point of greatest deflection, is at five-ninths of the length from the fixed end, and the point of inflection, or reversed curvature is about one-tenth the length from the fixed end.

When such a beam is loaded uniformly, Fig. 12, its greatest deflection is one-third of that of the same beam, or similar one of the same length, supported at both ends and loaded in the middle; or eight-fifteenths of that when loaded uniformly. The point of greatest deflection is at five-eighths the length from the fixed end, and the point of reversed curvature at oue-fourth the length from the fixed end.

Fig. 12.

40. *When the beam is fixed at both ends.* When such a beam is loaded in the middle, Fig. 13, its deflection is one-fourth of that of a similar beam merely supported at both ends. The point of reversed curvature is at one-quarter l from the end.

Fig. 13.

If the beam is loaded uniformly, Fig. 14, its deflection is only one-fifth as much as when supported at both ends and loaded uniformly, or one-eighth as much as when supported at both ends and loaded in the middle. The point of reversed curvature is $\frac{211}{1000} l$ from the end.

Fig. 14.

When a beam extends over several supports the outer portions are in the condition of Art. 39, and the inner portions is that of Art. 40. Such is the case of a bridge of several spans which are connected together over the piers. Then the middle portions may be made longer than the others.

41. *A beam supported at several points.* When a beam is uniformly loaded and is supported at several equi-distant points at the same level, different points of the beam are differently strained, being in the conditions given in Arts. 39 and 40, and the pressure upon the supports is not equally distributed. Thus if the uniformly loaded beam in Fig. 15, were cut in two at the point, C, that post would support one-half the load, and A and B each one-quarter; but if the beam be continuous C would support five-eighths of W, and A and B each three-sixteenths.

Fig. 15.

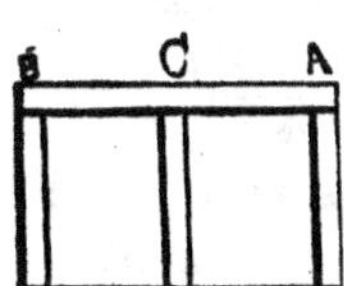

The same takes place in the tie beam of a roof truss. Similar results occur with any number of supports. Thus with two intermediate supports, the two middle posts each support eleven-thirtieths W, and the end ones each four-thirtieths W.

The supports may be made proportionally strong, or the pressure on them may be equalized, by a slight change in their relative weights.

Flexure of Rectangular Beams.

42. The general formula for prismatic beams for a beam supported at one end and loaded at the other, was

$$\delta = \frac{W l^3}{3 E I}. \quad [3.]$$

Let b = breadth of a rectangular beam and h = height or depth. Then $I = \frac{b h^3}{12}$. (Jackson's Mechanics, par

49.) Substituting this in [3] we get $\delta = \frac{4\,W l^3}{E b h^3}$. (7).

This shows that the deflection is directly as the weight, and as the cube of the length, and inversely as the breadth, and the cube of the height or depth. The ratios of deflection in other conditions of beams are the same as those of prismatic beams in general.

The most convenient standard of comparison is the important case of beams supported at both ends, and loaded in the middle, for which we have $\delta = \frac{W l^3}{4 E b h^3}$ [8.]

All the above formulas suppose the same unit of dimension to be used. It is, however, most convenient in practice to take the length in feet, and the breadth and height in inches, and to obtain the deflection in inches. The above formula [8] may then be written $\delta = \frac{W L^3}{e b h^3}$. In which L = length in feet, and $e = E \div 432$. Since

$$\delta = \frac{W l^3}{4\,E b h^3} = \frac{W\,(12\,L)^3}{4\,E b\,h^3} = \frac{W L^3}{E \div 432\,b\,h^3} = \frac{W L^3}{e\,b\,h^3}. \quad [9.]$$

Taking formula [9] for the standard, the deflections for the other cases are found by the ratios given in the case of prismatic beams in general. (See Art. 72)

Flexure of Prismatic Beams of other Forms.

43. To determine this, substitute in the general expression of the moment of flexure, $\delta = \frac{W l^3}{3\,E\,I}$ [3], the proper value of I for the particular cross-section in question. The following are the most useful.

For a hollow beam of a uniform rectangular cross-section, the moment of inertia, I equals $\frac{b h^3 - b' h'^3}{12}$; in which b' and h' are the inside breadth and height.

Fig. 16.

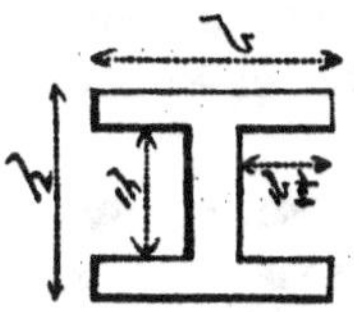

For an I shaped beam with dimensions as in the Fig. 16, the value of I is the same, and the preceeding formulas apply without change, and as before $I = \frac{b h^3 - b^1 h^{13}}{12}$. The two are therefore equally stiff with the same amount of material. For a square beam with its diagonals vertical and horizontal, $I = \frac{1}{12} c^4$, c being one of the sides of cross-section. This is the same as that for a square beam with its side horizontal and vertical: hence each beam has the same resistance to flexure. (N. B. It is not so for the resistance to rupture.) For a cylindrical beam of radius r, $I = \frac{1}{4} \pi r^4$.

For a square beam whose section circumscribes the section of a circular beam $I = \frac{1}{12} (2 r)^4$, therefore its resistance to flexure is to that of the circular beam $:: \frac{1}{12} (2 r)^4 : \frac{1}{4} \pi r^4 :: 1 : 0.59$.

44. Relative deflection of similar beams, or those having an equal cross-section.

Figs. 17 and 18.

In Fig. 17, $I = \frac{1}{12} [b (n b)^3 - b_1 (n b_1)^3]$.
$= \frac{1}{12} (n^3 b^4 - n^3 b_1 4)$.

In Fig. 18, $I' = \frac{1}{12} b_{11} (n^3 b_{11}^3)$.
$= \frac{1}{12} n^3 b_{11}^4$.

Now $n b_{11}^2 = n b^2 - n b_1^2$;

therefore $b_{11}^2 = b_2 - b_1^2$; therefore $I' = \frac{1}{12} n^3 (b^2 - b_1^2)^2$.
Then $I : I' :: b^4 - b_1^4 : (b^2 - b_1^2)^2$; $I : I' :: b_2 + b_1^2 : b^2 - b_1^2$.

(The letters on Figs. 17 and 18 are reversed.)

General Principles of Rupture.

45. The resistance to rupture or breaking, is examined in the same way as that to flexure or bending. Let R = resistance or strain, just before giving way, of the fibre most extended or most compressed. Call the distance from neutral axis v; and let a = area of the cross-section of the body in question.

Then its resistance $= R\, v\, d\, a$, and the resistance of any other fibre at a distance v_1 from the neutral axis $= R \times \frac{v_1\, d\, a.}{v}$ Its moment $= R \times \frac{v_1^2\, d\, a}{v}$. The total moment of the resistance of all the fibres equals the integral of the above from 0 to v on both sides of the neutral axis =

$$\int_0^v R \frac{v_1^2}{v} d\, a. \quad [10.]$$

Equating this in with the moment of the bending force, tending to produce rupture, we can determine the proper dimensons, &c.

Rupture of Prismatic Beams.

46. For these the total moment of resistance to rupture becomes $= \frac{R}{v} \times I.$ [11.]

Beams fixed at one end.

47. When loaded at the other end with W, the shearing or vertical force is everywhere W. (A force acting perpendicularly to the beam in any position may be substituted for W.) The bending or horizontal moment at any distance x, from fixed end, length being l, is $W\,(l-x)$, and at fixed end is therefore $W\,l$.

The equation then is $W l = \frac{R I.}{v}$ [12.]

This is the dangerous section of the beam. From this equation we can determine W, the breaking weight of any given beam; also I, which gives the size of the beam to resist any given stress W, also, R, or the coefficient of strength, having found W by direct experiment.

When the force is applied at any intermediate point of the beam, formula [12] applies, calling l the distance to the fixed end.

48. *When the beam is uniformly loaded with $w l$.* Then the shearing force at a distance x from fixed end is $w (l - x)$, and at the fixed end $= w l$. The bending moment at x is $w (l - x) \frac{1}{2} (l - x) = \frac{1}{2} w (l - x)^2$, and at the fixed end (the "dangerous section") is $\frac{1}{2} w l^2 =$

$$\frac{R I}{v}. \quad [13.]$$

That is, the stress is one-half as much as before. The weight of the beam itself acts as a uniform load.

Beams supported at both ends.

49. *When loaded in the middle with W.* Then the shearing stress everywhere is $\frac{1}{2} W$. Calling x the distance from the middle the bending moment at that point $= \frac{1}{2} W (\frac{1}{2} l - x)$. Calling x_1 the distance from the nearest end of the beam this becomes $\frac{1}{2} W x_1$. It is therefore greatest when $x = 0$; *i. e*, at the middle of the beam where it is

$$\frac{1}{4} W l = \frac{R I}{v}. \quad [14].$$

Hence it will require four times the weight to break it.

At the end of the beam it is equal to 0. The shearing force being everywhere $\frac{1}{2}$ W, may be represented by the perpendiculars of a rectangle whose length equals l, and whose height $= \frac{1}{2}$ W. The bending moment may be represented by the perpendiculars of an isosceles triangle whose base equals l and whose altitude equals the bending force in the middle of the beam.

50. When a beam, supported at both ends, is loaded, uniformly with $w\ l$. Then the shearing force at any point x from the middle $= w\ x$. It is therefore 0 in the middle and greatest at the ends of the beam, where it becomes $\frac{1}{2}\ w\ l$.

It may therefore be represented by the perpendiculars as in Fig. 19. The bending moment at any point x from the middle $= \frac{1}{2}\ w\ (\frac{1}{2}\ l + x)\ (\frac{1}{2}\ l - x)$. Calling x_1 distance from nearest end this becomes $\frac{1}{2}\ w\ x_1\ (l - x_1)$.

Figs. 19 and 20.

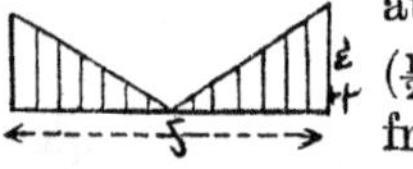

Proof.—Fig. 20. Reaction of $A = \frac{1}{2}\ w\ l$ upwards, with leverage x_1; pressure downwards of weight from A to $M = w\ x_1$ with leverage equal to $\frac{1}{2}\ x_1$. Resulting moment around $M = \frac{1}{2}\ w\ l\ x_1 - \frac{1}{2}\ w\ x_1^2 = \frac{1}{2}\ w\ x_1\ (l - x_1)$.

It may therefore be expressed by the perpendiculars to the double ordinate, terminated by a parabola. Fig. 21.

The bending moment is greatest at the middle $= \frac{1}{8}\ w\ l^2$

$$= \frac{1}{8}\ W l^2 = \frac{R\ I}{v}. \quad [15.]$$

To prove that the curve is a parabola, Fig. 22, $y = \frac{1}{8}$

$w l^2 - \frac{1}{2} w (\frac{1}{2} l + x)(\frac{1}{2} l - x) = \frac{1}{8} w l^2 - \frac{1}{8} w l^2 + \frac{1}{2} w x^2 = \frac{1}{2} w x^2$; or changing the letters $x = \frac{1}{2} w y^2$, the equation of the parbola.

Figs. 21 and 22.

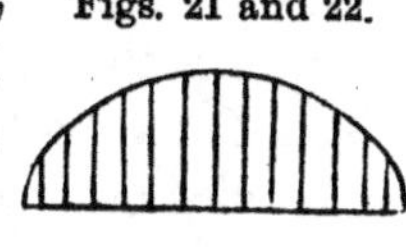

These principles apply to trussed bridges, which are framed beams, and show that those parts of them which have to resist a vertical stress, as do the posts and braces, should increase in size from the middle of the bridge to its ends in the uniform ratio shown in Fig. 19, and that those parts which have to resist the horizontal bending, as the top and bottom beams or chords, should decrease in size from the middle to the ends in the ratio shown by the perpendiculars in Fig. 21.

51. The lines of principal stress in beams uniformly loaded, are curved lines, such that the tangents to them at any points indicate the directions at these points of the lines of stress. Thay all intersect each other at right angles, and make angles of 45° with the neutral axis, on which line or plane, therefore, forces are acting in those directions. The lines convex upwards are lines of thrust, those convex downwards are lines of tension. The stress along each of these lines is greatest where it is horizontal, and at the end of each the stress is 0.

To show this experimentally, draw a number of small circles on the side of a beam before it is loaded. Then when it is loaded the circles will become ellipses. Those near the top of the beam will have their long axes vertical, and their short ones horizontal; those near the bottom of the beam, *vice versa;* and along the neutral axis these axes will make angles of 45° with the horizon. At intermediate points their figures will be intermediate in form.

52. *A beam loaded with W at distances l_1 and l_2 from the ends, l_1 being the smaller.* The shearing force between W and the nearest end is $\frac{l_2}{l} \times W$, on the other side of W, it is $\frac{l_1}{l} \times W$. The corresponding bending moments at any point distant x from the centre of the beam, are $\frac{l_2}{l}(\frac{1}{2} l - x) W$, and $\frac{l_1}{l} (\frac{1}{2} l + x) W$. x measured to the left of the middle is considered positive; to the right negative. The greatest moment is where W is applied, where it is

$$\frac{l_1 l_2}{l} \times W = \frac{R\,I}{v} \quad [16.]$$

In words this bending moment $=$ W multiplied by the product of the two parts of the beam, divided by its length.

Calling x_1 the distance from the middle at which W is applied, the shearing force for points between W and nearest and fartherest ends of beam are respectively

$$\frac{\frac{1}{2} l + x_1}{l} \text{ and } \frac{\frac{1}{2} l - x_1}{l} \times W.$$

The corresponding bending moments at distance x_1 from the middle, for points between W and nearest end are $\frac{(\frac{1}{2} l + x_1)(\frac{1}{2} l - x)}{l} \times W$, and on the other side of W, $\frac{(\frac{1}{2} l - x_1)(\frac{1}{2} l + x)}{l} W$. x is positive or negative according as it is to the left or right of the middle. The greatest bending moment is

$$\frac{(\frac{1}{2}\,l - x_1)\ (\frac{1}{2}\,l + x_1)}{l}\ W = \frac{R\,I}{v}.\ [17.]$$

In the first half of the above formulas, the point of application of W, is measured from the ends of the beam, and in the latter half from the middle. In the former l_1 and l_2 are used; in the latter these distances equal $\frac{1}{2}\,l - x_1$, and $\frac{1}{2}\,l + x_1$.

53. *With two forces equal, and at distances each equal to a from the ends of the beam.* The strength of such a beam is the same as if both forces were applied at the middle of a beam of length $= 2\,a$.

54. *With a weight, W, uniformly distributed over a length m and its centre of gravity in the middle of the beam.* The greatest bending moment is $\frac{1}{2}\,W \times \frac{1}{2}\,l - \frac{1}{2}\,W \times \frac{1}{4}\,m = \frac{1}{4}$

$$W\,(l - \tfrac{1}{2}\,m) = \frac{R\,I}{v}.\ [18.]$$

Therefore the bending moment equals that of a force acting in the middle $= W\left(\frac{l - \frac{1}{2}\,m}{l}\right)$.

If $m = l$ then formula becomes $\frac{1}{8}\,W\,l$.
If $m = 0$ then formula becomes $\frac{1}{4}\,W\,l$.

With the same load having its centre of gravity at distances l_1 and l_2 from the ends, the greatest bending moment is

$$W\frac{l_2 \times l_1}{l} - \tfrac{1}{2}\,W \times \tfrac{1}{4}\,m = W\left(\frac{l_1 \times l_2}{l} - \frac{m}{8}\right) = \frac{R\,I}{v}\ [19]$$

Partially loaded beams.

55. For a given intensity of load per unit of length, a uniform load over the whole beam produces a greater bending moment than any partial load. Consequently the removal of a load from any portion of a beam diminishes the bending moment at any point. For a given intensity of load per unit of length, the greatest shearing force at any given cross section of a beam, takes place when the longer of the two parts on either side of that point is loaded and the shorter is unloaded. The shearing force at any cross-section, at x distance from the middle, Fig. 23, of the beam, l being its length, and the longer part $A\,D = \frac{1}{2}\,l + x$ being loaded with w on each unit of length is $w\,\frac{(\frac{1}{2}\,l + x)^2}{2\,l}$. The shearing force at D for a uniform load over the whole beam is $\frac{1}{2}\,w\,x$. Then the excess of the shearing force for the partial load above that of a uniform load is $w\,\frac{(\frac{1}{2}\,l - x)^2}{2\,l}$. At the middle of the beam half loaded and half unloaded, this excess is at its maximum, for $x = 0$ and its whole shearing force equals $w\frac{(\frac{1}{2}\,l)^2}{2\,l} = \frac{1}{8}\,w\,l$. At the end of the beam this excess becomes 0. These points are important in proportioning the posts, ties and strees of frame bridges. With a load over the whole bridge, it stands; take off one-half and it breaks down.

Fig. 23.

56. *A beam fixed at one end and supported at the other.* When loaded in the middle this beam will support one

and one-third times as great a weight as a beam merely supported at both ends.

When uniformly loaded it has only the same strength as a beam supported at both ends. The place of the maximum stress is, however, now changed from the middle of the beam to a distance of five-eighths of the length from the fixed end.

57. *A beam fixed at both ends.* When loaded in the middle it is twice as strong as if it were merely supported at both ends. When uniformly loaded it is one and one-half times as strong as if it were merely supported at both ends.

58. *A beam extending over several supports.* Fig. 24.

Fig. 24.

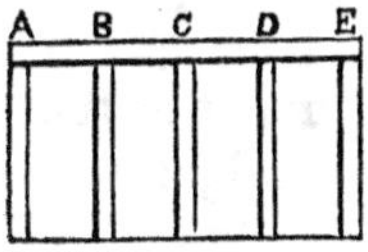

The inner portions as $B\ C$, $C\ D$, &c., between the supports are approximately fixed at both ends and are therefore for, a uniform load one and one-half times as strong as a separate beam. The extreme portions of the beam, $A\ B$ and $D\ E$, in the Fig., are beams fixed at one end and supported at the other, and therefore for a uniform load are no stronger than a separate beam.

Let a separate beam of a length l have a certain strength. Then the intermediate portions, $B\ C$ and $C\ D$, as in the last Fig., will have the same strength, if made of a length $= l \sqrt{1\frac{1}{2}} = l\ (1,225)$. Otherwise expressed (as obtained from the formula for the deflection of different parts), the length of ending span is to length of intermediate span $: : 1 : 1\frac{1}{4}$ about.

These principles are important in their application to proportioning bridges built in continuous spans.

Limiting length of beams.

59. The load which any two similar beams can bear, varies as the squares of any two corresponding dimensions; but their weights vary as the cubes of these. Therefore, the weights increase faster than the strength, and consequently there must be some limit of length at which a beam could not bear its own weight. This can be calculated, calling the weight of the beam a load uniformly distributed.

Let w_1 = weight of material per cubic feet, then $W = w_1\left(\frac{b}{12} \times \frac{h}{12} \times L.\right)$

If supported at both ends,

$$W = \frac{w_1\, b\, h\, L}{12 \times 12} = 2\, S \times \frac{b\, h^2}{L} \therefore L = 12\, \sqrt{\frac{2\, s\, h}{w_1}}.$$

This shows that the limiting length is independent of the breadth, but varies as the square root of the height. If supported at one end, $L = 6\, \sqrt{\frac{2\, s\, h}{w_1}}$, or ½ the former length.

Rupture of Rectangular beams with sides horizontal and vertical.

Beams fixed at one end.

60. When loaded at the other end with W, then the general equation [12] for prismatic beams is $W\, l = \frac{R\, I.}{v}$

For a rectangular section we have $I = \frac{1}{12}\, b\, h^3$ and we will take $v = \frac{1}{2}\, h$. Substituting we obtain

$$W = R\left(\frac{b\,h^2}{6\,l}\right) \quad [20.]$$

It is usual to take the length L in feet, and b and h in inches. Then the above expression becomes $W = R \left(\frac{b\,h^2}{72\,L}\right)$. It is also usual to employ a new coefficient $s = \frac{1}{18}\,R$, for reasons given under the next head. Then we have finally $W = \frac{s\,b\,h^2}{4\,L}$ [21].

When the same beam is loaded uniformly with $w\,l$ we have $w\,l = \frac{s\,b\,h^2}{2\,L}$ [22.]

Beams supported at both ends.

61. This contains the most important cases in practice.

When loaded in the middle. Then $W = \frac{s\,b\,h^2}{L}$. [23.]

Putting b, h, and L each equal to unity we get $W = s$. Then s may be defined as the weight which applied at the middle of a beam, 1 inch square, 1 foot long, between supports, would just break it, neglecting its own weight. It should be determined by experiment, for reasons to be given under Art. 76. This is the reason why s is used instead of $\frac{1}{18}\,R$.

The above formula [23] may also be attained by very simple reasoning. The strength of a beam is directly as the number of fibres it contains, or as its cross-section; i. e., for a rectangular beam as $b \times h$; and the resistance of each of these is as its distance from the neutral axis, and therefore as the height of the beam or as h. The leverage of the weight is as the length of the beam. Therefore its strength is directly as $b \times h \times h$ and inversely as l; i. e., as $\frac{b h^2}{l}$, and also as a certain constant coefficient s, depending on the material.

When the beam is uniformly loaded with $w L$, then we have $w L = 2 s \times \frac{b h^2}{L}$ [24.]

62. When the beam is loaded at any other point, distant L_1 and L_2 from the ends. then we have $W = \frac{1}{4} s b h^2 \times \frac{L}{L_1 L_2}$ [25.]

63. With two forces equal and at equal distances a from the ends. Denoting the sum of the two forces by W, we have $W = s \times \frac{b h^2}{2 a}$.

64. When the weight is uniformly distributed over m, and the centre of m is over the middle of the beam. In this we have $W = s \times \frac{b h^2}{L - \frac{1}{2} m}$ [26.]

65. When the weight is distributed over m, and the centre of m is at distances L_1 and L_2 from the ends of the beam, L_1 being the nearer. In this we have

$$W = \frac{1}{4} s \times \frac{b h^2}{\frac{L_1 + L_2}{L} - \frac{m}{8}}$$

66. Partially loaded beams; same as Art. 55.

67. A rectangular beam fixed at one end and supported at the other; same as Art. 56.

68. A rectangular beam fixed at both ends; same as Art. 57.

69. A rectangular beam extending over several supports; same as Art. 58.

RUPTURE OF PRISMATIC BEAMS OF OTHER FORMS THAN RECTANGULAR.

70. The general expression for equation of rupture is $\frac{R I}{v}$ = moment of the force applied. For a rectangular beam supported at both ends and loaded in the middle $W = \frac{2}{3} R \left(\frac{b h^2}{l}\right) = s \times \frac{b h^2}{L}$. [23.] For a square beam, its sides being c, $W = s \times \frac{c^3}{L}$. For the same beam placed diagonally, $W = s \times \frac{c^3}{L \sqrt{2}}$.

For a hollow, rectangular beam, $W = s \times \frac{b h^2 - b_1 h_1^2}{L}$, b and h, and b_1 and h_1 being respectively the outside and inside breadths und heights. Compared with a solid beam containing equal material, the cross-sections being similar rectangles, and the sides of the hollow beam being n times those of the solid beam, their relative strengths are; strength of hollow : strength of solid beam :: $2 n - \frac{1}{n}$: 1. Thus if n be 10, the relative strengths are :: $20 - \frac{1}{10}$: 1 or :: 19. 9 : 1.

The I shaped beam has the same strength as the preceding one.

For a cylindrical beam with radius r, we have $I = \frac{1}{4}\pi r^4$, $v = r \therefore W = \frac{R \pi r^3}{4 l} = \frac{3}{8} s \times \frac{\pi r^3}{L}$. For a beam fixed at one end and loaded at the other; or for a beam supported at both ends and loaded in the middle

$$W = \frac{3}{2} s \times \frac{\pi r^3}{L}. \quad [28.]$$

Comparing the round beam with its circumscribing square beam, whose side $= 2r$ we find the ratio to be as .6 : 1. Compared with a square beam with the same quantity of material, since $\pi r^2 = c^2$; the strength of the square beam : strength of round $:: s \times \frac{c^3}{L} : \frac{3}{2} s \frac{\pi r^3}{L} :: s \frac{(r \sqrt{\pi})^3}{L} : \frac{3}{2} s \frac{\pi r^3}{L} :: 1.18 : 1 :: 1 : .847$.

The strength of a hollow, cylindrical beam equals the difference of the strengths of the inner and outer sections. $W = \frac{3}{2} s \pi \frac{(r^3 - r_1^3)}{L}$, r beiug the outer and r_1 the inner radius. Comparing it with a solid cylindrical beam of this same material whose radius is $\frac{1}{n}$ of the hollow beam, we find; strength of hollow beam : strength of solid beam $:: 2n - \frac{1}{n} : 1$.

Forms of Iron Beams.

71. In some materials the resistance to extension and compression are very different. Hence beams whose tops and bottoms are unlike will vary in strength with their

position. Thus a cast iron beam of T form, broke with 2½ cwt., being placed on supports with the flange upwards. A similar beam broke with 9 cwt. when placed flange downward. This is because cast iron resists compression much more than it resists extension, and a beam of it will therefore resist more, when more of the material is placed in the extended parts, so that the greater quantity of material placed there may compensate for its less power to resist the kind of stress acting on it.

The strongest form of a beam is therefore one in which most of the material is placed as far as possible from the neutral axis; that is hollow or I shaped, and also so disposed that the portions resisting compression and extension shall be in quantities inversely proportional to their powers of resisting these stresses.

This ratio for cast iron is 6 ; 1. Therefore the strongest form for it, is that in which the lower flange is six times the upper. This is known as Hodgkinson's Beam. Such a beam is about one-third stronger than the I shaped beam, containing the same amount of material. In practice the upper flange is made somewhat larger than this ratio to prevent twisting. In such beams the strength is very nearly directly as the area, a, of the bottom flange, and as the depth, h, of the whole beam, and inversely as the length, L.

The formula for the breaking weight applied in the middle of a beam supported at both ends is

$$W = 2\tfrac{1}{6}\,\frac{a\,h}{L}. \quad [29.]$$

In which W is in gross tons, h in inches, a in square

inches, and L in feet. In this formula the weight of beam is taken in. Thus such a girder 30′ long, 2′ high, and the bottom flange 20″ × 2″, would break with a weight in the middle equal to $\frac{2\frac{1}{6} \times 40 \times 24}{30} = 69$ tons gross. The greatest load should never be more than one-third the breaking weight, or one-sixth of it if the beam be subjected to vibrations as in railroad bridges. Such beams are proved by applying weights, or by an hydraulic press. This proof strain should not be more than one-half that which would break them. Their strength is inferred during the proof, from the fact that the breaking weight is almost twice that which causes a deflection of $\frac{1}{480}$ of its length. Thus in the above beam, 34½ tons should cause a deflection of three-fourths of an inch.

If, however, the beam is to be subjected only to small stresses, not approximating rupture, the top and bottom should have the same cross-section, since its resistance under such stresses are about the same to compression and extension.

72. Table, showing the relative deflections, and breaking weights, of beams having the same cross-section and length, and placed in different conditions. A beam, supported at both ends and loaded in the middle, is taken as the standard. δ is its deflection and W its breaking weight:

BEAMS.	Figs.	Deflection. Load constant.	Breaking weight.
Supported at both ends and loaded in the middle, - - - -	9	δ	W
Supported at both ends and loaded uniformly, - - - -	10	$\frac{5}{8}\,\delta$	$2\,W$
Fixed at one end and loaded at the other, - - - - -	7	$16\,\delta$	$\frac{1}{4}\,W$
Fixed at one end and loaded uniformly, - - - - -	8	$6\,\delta$	$\frac{1}{2}\,W$
Fixed at one end, supported at the other and loaded in the middle, -	11	$\frac{45}{100}\,\delta$	$1\frac{1}{3}\,W$
Fixed at one end, supported at the other and loaded uniformly, - -	12	$\frac{1}{3}\,\delta$	$2\,W$
Fixed at both ends and loaded in the middle, - - - - -	13	$\frac{1}{4}\,\delta$	$2\,W$
Fixed at both ends and loaded uniformly, - - - - -	14	$\frac{1}{8}\,\delta$	$3\,W$

For a rectangular beam with sides horizontal and vertical, when supported at both ends and loaded in the middle, $\delta = \frac{W\,L^3}{e\,b\,h^3}$. [See Art. 42); in which $W =$ the weight applied, $L =$ length of beam in feet, b and $h =$ breadth and height in inches, and $e = E$, divided by 432. E is found by experiment for different materials. (See Art. 15 and tables in Arts. 18 and 19.) For the same beam $W = s\,\frac{b\,h^2}{L}$. (See Art. 61); in which b and $h =$ breadth and height in inches, $L =$ length in feet, and s is a coefficient found by experiment. (See Art. 76, and tables).

73. *Relative strength of cast iron beams.* Fig. 25 (a) is a cross-section of a beam made by Boulton & Watts in 1801. It was improved by Fairbairn in 1825, the vertical rib being made thinner and the lower flange thicker (b). Tredgold's beam (c) has equal upper and lower flanges. The strongest form is Hodgkinson's (d), the lower flange being six times the upper one. The relative strength of these beams, Hodgkinson's being taken at unity, is: Boulton & Watts', 0.51; Fairbairn's, 0.75; Tredgold's, 0.62, and Hodgkinson's, 1.

Fig. 25.

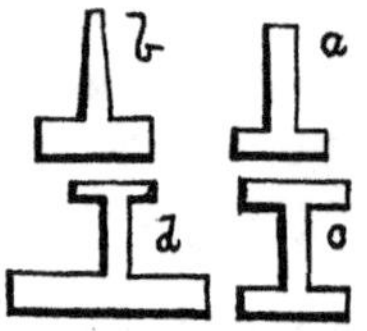

74. Cast iron beams sometimes have wrought iron tension rods applied to them with the object of strengthening the lower flange; the two rods helping it to resist extension. The rods are tightened by screws or wedges, so as to have any amount of initial tension in advance; but it is difficult so to adjust the two so that each shall bear its share of the strain; and even if this adjustment were once made it would be altered after any strain, owing to the different "*sets*" of wrought iron and cast iron; since for respective stresses equal to two-thirds breaking weight, for each (say five tons per square inch for cast iron and fifteen tons for wrought) the elongation for wrought is two and one-half times that of cast and its *set*, ten times as great as that of cast. This adjustment, and with it the strains coming on each, would also vary with every change of temperature, since wrought iron expands with heat more than cast iron. The combination is therefore bad.

Wrought iron resists extension much more than compression, therefore the compressed parts of wrought iron beams (the upper flange of a beam supported at both ends) should be nearly as 2 : 1. They are usually made nearly the same, since for small strains its resistances are about the same.

Fig. 26.

A common form is this. Fig. 26. Usual ratio of depth to span about 1 : 14.

The box or tubular form, Fig. 27, is stiffer but less easily kept free from oxidating, and is more difficult of construction.

Fig. 27.

Wood resists extension about twice as much as compression, therefore where it is formed into compound beams, such as the sides of bridges, the lower string pieces need be only one-half as large as the upper ones, if it were not for the weakness, produced in the portions of beams subjected to extension, by any defect in the material, such as knots or decay, and by the notches and holes necessary for connecting the parts together.

Experiments on Rupture.

76. $R =$ the resistance of the material under discussion to yielding by extension, which has already been given. But direct experiments of the resistance of beams to breaking across give a greater value for R in the case of breaking. This shows that some other element of strength exists.

Barlow Experiments. (Civil Engineering Journal, 1856,

page 9.) These prove an additional resistance proceeding from the lateral action of the fibres or particles resisting sliding on one another as they must do where a beam is bent, and increasing with the depth of the beam. The value of R, to be used in formulas for transverse strength, must therefore be obtained directly.

The most convenient unit will be the weight in pounds which would break a beam or bar 1″ square and 1′ long, when applied to its middle. This will be expressed by s. It has been shown that $R = 18\,s$, which is the value of R to be used in the subsequent formulas. The following values of s are obtained from direct experiment:

For cast iron $s = 1700$ to 2600.

For wrought iron $s = 2300$ to 3000.

Safe load from one-quarter to one-sixth breaking weight.

Table for American Timber.

NAME.	s.
Ash,	600
Ash, Black,	290
Ash, Swamp,	390
Beech, White,	460
Beech, Red,	580
Birch, Black,	700
Birch, Yellow,	440
Cedar, White,	255
Hemlock,	380
Hickory, Bitternut,	490
Hickory,	700
Oak, Live,	600
Oak, Red,	560
Oak, White,	580
Pine, Red,	510
Pine, Yellow,	395

Name.	s.
Pine, White,	410
Pine, Pitch,	580
Spruce,	340
Tamarack,	300

The safe stress for timber is variously taken at from one-fourth to one-tenth the breaking weight, depending on the use to which it is to be put and cautiousness of the builder.

Transverse Strength of American Timber.

Name.	$s_1 = \frac{W l}{4 b d^2}$	$s = \frac{W L}{b d^2}$
Hickory,	2,129	710
Birch, Black,	2,061	687
Oak, Live,	1,862	621
Ash,	1,795	598
Oak, White,	1,743	580
Beach, Red,	1,739	580
Pine, Pitch,	1,727	576
Oak, Red,	1,687	562
Pine, Red,	1,527	509
Hickory, bitternut,	1,465	488
Beech, White,	1,380	460
Birch, Yellow,	1,335	445
Pine, White,	1,229	410
Pine, Yellow,	1,185	395
Ash, Swamp,	1,165	388
Hemlock,	1,142	380
Spruce,	1,036	345
Tamarack.	911	304
Ash, Black,	861	287
Cedar, White,	766	255

American Building Stone.

NAME.	*s*.
Blue stone flagging,	125
Quincy granite,	104
Blue granite (coarse-grain),	72
New Jersey freestone,	71 to 96
Connecticut "	52
Dorchester "	43
Caen-stone,	24

Safe stress about one-tenth

Solids of equal Resistance.

77. These are solids of such a form that they are equally strong in every part of their length. Such forms are the strongest possible for any given amount of material. For if the beam be of any other form so that any part be about to give way then from the first part may be taken away a portion of material and added to the second part, thus rendering the beam stronger than before. Therefore the strongest form of a beam is that in which there is an equal liability to rupture in every point.

A beam fixed at one end and loaded at the other. If the depth be uniform the plan must be a triangle; for in the expression for strength, at any point x from the end $W = \frac{1}{4} s \left(\frac{b\,h^2}{x}\right)$, in which all the quantities are constant except b and x therefore b varies as x, and the plan is a triangle.

If the beam is to have a uniform breadth its elevation must be a parabola or one-half a parabola. For in the general equations b being constant, h^2 must vary as x; i. e., the abscissas vary as the squares of the ordinates, which is a property of the parabola. This is therefore the proper form for a bracket supporting heavy weights at its end. Its depression is double that of a prism.

If the beam be supported in the middle and loaded equally at both ends, each half of it is in the same condition as the preceding beam, and its form would be that of two such beams placed end to end, i. e., two whole or half parabolas. Such is the form given to beams of balances and the walking beams of steam engines.

This form is often modified into a straight-lined figure, including the parabola, its upper and lower sides being tangent to the parabola at the fixed points, and by making the depth of the beams at the ends one-half that at the middle (which is determined as above). One-third is sometimes used instead of one-half, as an economical approximation.

If the cross-sections are required to be similar, the breadth and height vary as the cube root of the levers arm $\left(\frac{x}{l}\right)$. For a uniform load and similar cross-sections both plan and profile must be a semi-cubical parabola, in which the cubes of the ordinates vary at the squares of the abscissas.

A beam fixed at one end and loaded uniformly. If the depth be uniform its plan will be formed by two ½ parabolas concave outwards. If its breadth be uniform its elevation will be a triangle. For a uniform load and similar cross-sections the profile and plan must be semi-cubical parabolas.

A beam fixed at one end and loaded at the other, and also

loaded uniformly. If its breadth be uniform its elevation will be an hyperbola, with the vertex at the outer end.

A beam fixed at one end and supporting its own weight only. Its breadth being uniform its profile will be a parabola concave downwards.

A beam supported at both ends and loaded at any point. If the depth be uniform the plan should be a pair of triangles joined by their bases at the point where the weight is applied. Sufficient width to resist the strain should, however, be added at the ends.

If the breadth be uniform the profile should be two parabolas meeting at the point where the weight is applied. Here too, provision must be made to resist the shearing force.

A beam supported at both ends and loaded uniformly. Where the depth is constant the plan should be two parabolas. Here also the ends should have enough material to resist the shearing force. If the breadth be constant the profile will be elliptical. For the strain at any point is proprtional to the product of the two parts, and the resistance is as the square of the depth, and these two things being proportional the curve thus formed is an ellipse. This form was formerly used for rails, but was abandoned from the inconvenience of laying them, and also because such a form, although the strongest, is less stiff than the corresponding rectangular form, as three to four.

Combining the principles of strongest cross-section and best longitudinal form, we find the strongest shape for a cast iron beam, uniformly loaded, to be two parabolas in plan, and Hodgkinson's form of cross-section. If the breadth be uniform and the load uniform, the profile is (approximately) a parabola concave downwards.

RESISTANCE OF POSTS, PILLARS, COLUMNS, STRUTS, &c., TO YIELDING BY FLEXURE.

General Principles.

78. They are supposed to be resisting a stress of compression applied in the direction of their length, either exactly or approximately through their axes. When their length is many times their diameter they give way, not by direct crushing, but by bending sidewise and by breaking across. The theory of this is very imperfect. One theoretical investigation would make $W = s_1\left(\frac{a\,t^2}{l^2}\right)$ [31]

or where round or square $W = S_{11}\left(\frac{t^4}{l^2}\right)$ [32.]

Another investigation is, for strength of a long pillar

$$W = \frac{S_{111}\,a}{1 + m\left(\frac{l^2}{t^2}\right)} \quad [33.]$$

in which W = breaking weight, a area of cross-section, l = length, t = thickness in the direction where it is least; all in same unit, (usually inches.) S_{111} = coefficient of strength, m = coefficient found by experiment. This is Gordon's Formula.

Stone Columns.

79. **Experiments make the strength of these to vary thus:**

Ratio of height to diameter, - -	1	12	24	30	40
Strength - - - - - - -	1	1	.70	.54	.37

Cast Iron Columns.

80. In formula [33], let all the dimensions be in the same unit. For cast iron let $S_{III} = 80{,}000$ pounds per square inch, $m = \frac{1}{400}$. Then it becomes

$$W = \frac{80{,}000 \times a}{1 + \frac{1}{400}\left(\frac{l^2}{t^2}\right)} \quad [34.]$$

This is Gordon's new formula.

Hodgkinson's formula, for the breaking weight of cast iron columns, their length being at least thirty times their diameter, is this; $W = 44\,\frac{d^{3.6}}{L^{1.7}}$ [35.]

L being the length in feet, d = diameter in inches, and W = breaking weight in gross tons. If W is required in pounds, instead of 44 use 100,000. A modification of the above formula is this, $W = S_{IV}\,\frac{d^{3.5}}{L^{1.63}}$ [36.]

In which S_{IV} varies from 50 to 33.6, the mean being 42.3.

This formula supposes the ends to be flat and firmly fixed. With rounded ends the strength is only one-third that of the former case. As columns are never perfectly fixed, one-sixth of the above breaking weight should be taken as a safe load. Where cast iron columns have a length less than thirty times their diameter, *Hodgkinson's Formula is* $W = \frac{b\,c}{b + \frac{3}{4}c}$. [37.]

In which b = breaking weight, given by formula [35], and c = the crushing weight for a short piece of the material.

When the above formula gives a strength greater than the resistance of the material to crushing, the latter should be taken as the true strength.

Ratio of length to diameter.	Breaking load in lbs. per sq. inch.	Ratio of length to diameter.	Breaking load in lbs. per sq. inch.
1	92,000	35	16,000
2 1-2	85,000	40	12,888
5	78,000	45	10,100
7 1-2	67,000	50	8,100
10	48,000	60	6,200
15	37,000	70	4,700
20	32,000	80	3,900
25	26,000	90	3,450
30	21,000	100	3,100

The breaking weight of a short piece varies greatly with the quality of the iron, say from 20,000 to 100,000 pounds. In the above table 92,000 is taken.

Capt. Meigs' practice at the Capitol.

Ratio of length to diameter.	Safe load for principal parts.	Safe loads for secondary parts.
$<$ 12:1	10,500	17,500
12:1	8,750	14,600
24:1	5,250	8,800
48:1	1,750	2,900
60:1	1,050	1,800

81. *For hollow cast iron pillars.* If d be the outer diameter and d_1 the inner diameter, the strength = the difference of strength of two solid columns of diameters d and d_1 and we have by Hodgkinson's formula W in tons $= 44\left(\frac{d^{3.6}-d_1^{3.6}}{L^{1.7}}\right)$ [38.]

Caution. In casting columns the core is sometimes not central, and therefore the sides of the column are of unequal thickness; such should be rejected. This was the probable cause of the fall of Pemberton Mills.

Mr. Francis would use only one-twelfth to one-fifteenth W for the safe permanent load.

82. *Wrought iron columns, &c.*

Hodgkinson's formula for wrought iron solid columns is $W = 134 \left[\frac{d^{3.55}}{L^2}\right]$ [39.] W and L representing the same as before, and d = side or diameter. For pounds the formula is $W = 300{,}000 \frac{d^{3.55}}{L^2}$.

Gordon's formula, [33] for wrought iron becomes

$$W = \frac{36{,}000\ a}{1 + \frac{1}{3000} \times \frac{l^2}{t^2}} \quad [40.]$$

By Gordon's formula we find;

Length : *d* or *t*		10	20	26½	30	40
Breaking load per square inch.	Cast iron,	64,000	40,000	29,000	25,000	16,000
	Wro't iron,	35,000	32,000	29,000	28,000	23,000

i. e., columns of cast iron are stronger than those of wrought iron up to a length equal to twenty-six times the diameter. For columns more slender wrought iron is the stronger.

Capt. MEIGS' experiments at the Capitol give the following results for safe strains per square inch for wrought iron:

Ratio length : diameter.	Safe strain.
$<$ 12:1	7,000
12:1	5,806
24:1	3,483
48:1	1,160
60:1	580

83. The strength of wrought iron plates acting as struts, varies approximately as their width and as the cube of their thickness, and inversely as the square of their height or length; corresponding to formulas [31] and [32.]

To calculate the strength of struts, having the form of angle iron, T iron or double T iron, apply Gordon's formula [40], calling t the thickness in the weakest direction; i. e., the direction in which it will give way, and considering separately the various parts into which it will be divided by planes parallel to that in which it would bend, for example, parallel to $A\ B$ in Fig. 28.

Fig. 28.

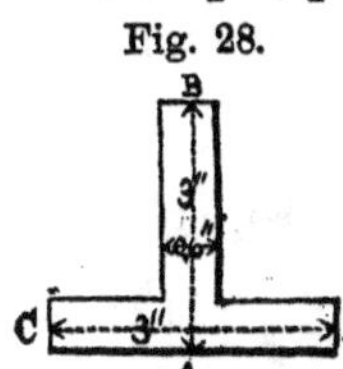

Example. Length 8′. For $A\,B\ a = 1.80$ square inches; for $C\,D\ a = 2.4'' \times .6'' = 1.44$ square inch, $t = 0.6''$. By Gordon's formula $W = 53{,}748$ pounds.

Actual breaking stress = 36,000. Using in the formula 24,000 instead of 36,000 would give the actual results. Experiments are wanting.

Experiments by N. Y. C. R. R. give results indicating in the iron used in this form the proper values of S to be only about 24,000. The strut of the example, of English crown iron behaved as follows in the experiments.

First experiment.

Load in lbs.	Flexure on *C D.*	Load in lbs.	Flexure on *C D.*
5,500	.0	22,000	.1875
8,200	.050	24,700	.220
10,900	.075	27,500	.250
13,700	.100	30,200	.312
16,500	.125		No set.

Second experiment.

Load in lbs.	Flexure on *C D.*
22,000	.126
30,400	.5125
36,000	Bar bent almost double.

Such experiments are very valuable and very scanty. For plate iron girders stiffened by T ribs, the depth across the ribs may be taken for t in Gordon's formula. For wrought iron cells, stiffened by angle iron, when the thickness of the plate is not less than $\frac{1}{30}$ of the diameter, the stress required to break them by bending is 27,000 pounds per square inch of the cross-section of the iron.

Resistance of wooden posts to breaking by flexure.

84. Hodgkinson's formula is this; $W = S\left(\frac{d^4}{L^2}\right)$ [41,]

which for Dantzic oak is $W = 24{,}500\left(\frac{d^4}{L^2}\right)$, and for red deal (pine) $W = 17{,}500\left[\frac{d^4}{L^2}\right]$, in which W equals breaking weight in pounds, d = side of square base in inches, and L = length in feet.

The crushing weight of the above woods are from 6,000

pounds to 10,000 pounds per square inch, for the former, and 7,000 for the latter.

Rondelet's experiments for wood crushing with 6,000 pounds per square inch, gave the following ratios of strength for long pieces:

Ratios of height to diameter or side, -	1	12	24	36	48	60	72
Ratio of resistance, - - - - -	1	$\frac{5}{6}$	$\frac{1}{2}$	$\frac{1}{3}$	$\frac{1}{6}$	$\frac{1}{12}$	$\frac{1}{24}$

He considers the safe permanent load to be $\frac{1}{7}$ of what the above table would give.

Mr. Whipple has also made experiments.

The safe loads according to the three above sets of experiments are given in the following table:

Table of safe load for Pillars, &c., in pounds per square inch.

Ratio of height to thickness,	1	12	14	16	18	20	22	24	28	32	36	40	48	60	72
Rondelet,	860	700						430			290		140	72	35
Whipple,	1,000	850	800	750	700	600	520	460	370	307	263	227	156	100	
Hodgkinson,								624	450	355	280	230	150	100	70

Table of safe mean of loads.

Ratio of height to thickness,	1	10	15	20	25	30	35	40	45	50	55	60	65	70
Safe load,	900	800	700	560	420	340	280	225	175	130	105	85	70	60

Resistance to Torsion.

85. Torsion is the twisting stress to which are subjected all parts of an axle to which are applied two forces, acting in contrary directions. It is opposed by the resistance of the material to extension and shearing.

The angular displacement of any particle of a twisted solid is proportional to its distance from the axis and also to the length of the solid, each fibre being bent into the form of a helix.

The resistance is as the fourth power of the diameter. Let α equal angle of torsion measured by parts of the radius (radius being equal to 57.3°), or arm of leverage of the power.

Let L equal the length in feet of the solid from the fixed end to the point where the force is applied. Let d = diameter of the solid and p = power in pounds acting to twist the solid. r_1 = its arm of leverage. L is in feet, d and r_1 in inches. Then for wrought iron we have

$$\alpha = \frac{p\, r_1\, L}{70000\, d^4} \quad [42.]$$

Multiplying α by the radius r_1 we get the absolute displacement.

For a square beam of side d, for wrought iron $\alpha =$

$$\frac{p\, r_1\, L}{116,000\, d^4} \quad [43.]$$

The stress which would break a body by torsion is found thus: For a cylindrical axle the moment of resistance to torsion is $\frac{1}{2}\, \pi\, s_1\, r^3 = 1.575\, s_1\, r^3 = .2\, s_1\, d^3$, of which s_1 = resistance to shearing, r = radius of axle in inches and d = diameter of axle in inches. Let p = power applied, and r_1 = its leverage. Then we have $.2\, s_1\, d^3 = p\, r_1$. Whence $d^3 = \frac{5\, p\, r_1}{s_1}$ [44.] (all in inches.)

For cast iron s_1 = 15,000 to 30,000, and for wrought iron s_1 = 50,000. Hence $d^3 = \frac{p\, r_1}{3,000}$ to $\frac{p\, r_1}{6,000}$ for cast iron, and $d^3 = \frac{p\, r_1}{10,000}$ for wrought iron.

For safety a small portion of s_1 must be used, depending on the strain.

Tredgold uses one-fourth for cast iron and one-sixth for wrought iron. Morin takes for wrought iron one-eighteenth for light strains and one thirty-sixth for great strain. Whildin uses one-sixth.

Morin gives the following formulas:

Material.	Great strain.	Light strain:
Cast iron	$d^3 = \frac{pr_1}{180}$	$d^3 = \frac{pr_1}{360}$
Wrought iron	$d^3 = \frac{p\,r_1}{540}$	$d^3 = \frac{p\,r_1}{1{,}080}$
Wood	$d^3 = \frac{p\,r_1}{30}$	$d^3 = \frac{p\,r_1}{60}$

One set of experiments makes the resistance to torsion of different metals as follows, taking *lead* for unity:

Lead,	1
Tin,	1 1-2
Copper,	4 1-3
Brass,	4 2-3
Hard gun metal,	5
Best wrought iron,	10
Blistered steel,	16 2-5
Shear steel,	17
Cast Steel,	19 1-2

Relative strength against torsion of cast iron shafts of various cross-sections and of equal area.—Major Wade:

Solid cylinder.	Solid square.	Hollow cylinders with the following ratios of interior and exterior diameter.				
		.4:1	.5:1	.6:1	.7:1	.8;1
1	.875	1.266	1.443	1.700	2.086	2.738

PART III.

The Stability of Position, or Resistance to Overturning.

Introduction.

86. A frame or structure may give way, either by its portions breaking, or by their being displaced The strength of the materials resists the former, the weight of the frame gives stability to the latter.

Stability is of two kinds. *The Stability of Position* is the resistance of a body to being displaced by turning over. *The Stability of Friction* is the resistance of a body to being displaced by sliding. These two kinds of stability constitute Parts III and IV.

Frames or structures may be classified in reference to the connection of their parts. When the joints are very narrow in proportion to the length of the pieces, as in roof trusses, &c., the combination may be called *Bar Work*. When the joints are wide, as in walls or arches, it may be called *Block Work*.

Bar Work.

Mechanical Principles.

(See Jackson's Mechanics; on representing forces by straight lines, on resultant forces, and ratios of forces being as the sines of opposite angles, &c., &c.)

87. When frames of three bars, (which frequently oc-

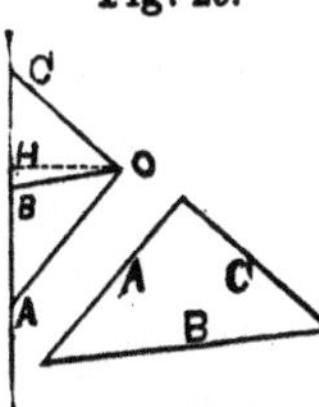

Fig. 29.

curs in the most important cases of engineering), are subjected to parallel forces, such as weights; the stress may be obtained thus: Let *A B* and *C*, Fig. 29, be three bars forming such a frame, with loads so applied as to be in equilibrium. From any point *O*, drawn parallels to the lines *A*, *B* and *C*, ending in a vertical line; then will the radiating lines, *O B*, *O A* and *C O* represent the respective strains on the bars to which they are parallel. The horizontal *O H* is, likewise, proportional to the horizontal thrust of the frame.

Elements.—Single pieces are the elements of frames. They may be subjected either to compression or extension. A compressed piece is called a "*strut.*" Its equilibrium is unstable; for if it move at all, the force *P* and the resistance *R*, tend to make it move farther. An extended piece is called a "*tie.*" Its equilibrium is stable; for when it is displaced, the forces *P* and *R* tend to bring it back again.

The term "*brace*" includes both "struts" and "ties." A tie must be in one piece, a strut may be in several. It

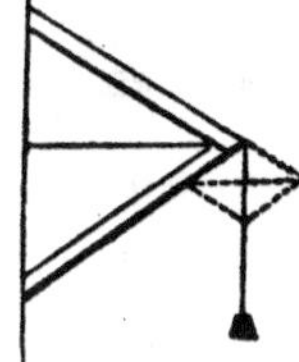
Fig. 30.

is often important to distinguish between ties and struts. If it is seen that a rope or chain may be substituted for a beam, it is a tie; otherwise it is a strut. Another and better way of distinguishing them is as follows: Construct a parallelogram on the straining force as a diagonal and with its sides parallel to the sustaining forces. Fig. 30. Draw the other diagonal of the parallelogram. Draw, parallel to this last diagonal, a line through the point where the directions of the forces meet. This line is the dividing line between struts and ties. Observe to which side the straining force would move if left at liberty. All supports on that side of the line would be compressed and will be struts. Those on the other side would be extended and will be ties.

The weight of a beam, or bar, is to be regarded as a vertical force acting at its centre of gravity. Let a uniform, straight beam rest with the extremities against smooth, horizontal and vertical surfaces, as in Fig 31. The beam is kept in equilibrium by its weight acting in the vertical through G, by the horizontal pressure h at B, acting in the line $B\,D$, and by a third force at A, which must act in the line $A\,D$. Since the triangle $A\,D\,E$ has its side parallel to directions of the three forces, these sides will be proportional to the forces, and we have $W : h :: D\,E : A\,E$, or $:: B\,F : \frac{1}{2}\,A\,F$. Hence $h = \frac{1}{2}\,W \times \frac{A\,F}{B\,F}$. By moments we have $W \times A\,E = h \times B\,F$ or $h = \frac{1}{2}\,W \times \frac{A\,F}{B\,F}$. Trigonometrically this becomes, $h = \frac{1}{2}\,W \times \text{cot.}\,B\,A\,F$. The thrust in the direction of the length of the beam $= \frac{1}{2}\,W \frac{A\,B}{B\,F} = \frac{1}{2}\,W \text{cosec.}\,B\,A\,F$.

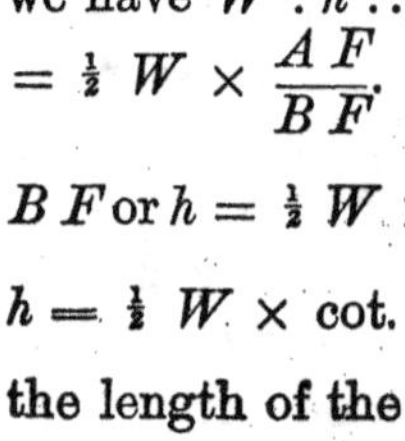

Fig. 31.

A pair of struts (single) Fig. 32.

88. The most important case of this is a roof truss. This is a frame supporting a load which lies between the points of support. Construct a parallelogram having for one diagonal a line representing the number of units in the weight, and having its sides parallel to the struts. These sides will represent the strains on the beams to which they are parallel. Any change in the obliquity of the beams increases, or diminishes the strain.

Fig. 32,

When the weight rests immediately on the struts, or is sustained above them, or is suspended below them, the effect is the same. The length of the beams has no effect on the stress, as is evident from the construction, though we have learned before that the strength of the beam diminishes as the length increases. The strain may also be calculated numerically; for the frame is kept in equilibrium by the action of three forces, viz; the weight and the resistance of each beam; therefore each force is proportional to the sine of the angle made by the directions of the other two. (See Jackson's Mechanics).

Calculation of Strains.

Fig. 33,

89. Let the weight = 1500 lbs. Let each inch of the diagonal represent 500 lbs. Let the angle of the beams = 100°, the one making an angle of 60° with the vertical, and the other an angle of 40°. Required the strain on the two beams, *A B* and *A C*. Fig. 33 (The Fig. is not drawn to scale).

1. *Graphically.*—Set off on the vertical 3″ and complete the parallelogram as before. The proportions of the weight born by B and C, are respectively $A\ G = D\ H$ and $A\ H = D\ G$. When B and C are at different heights, to find the portion of the weight borne by each, draw the lines, not horizontal, but parallel to $B\ C$.

2. *Trigonometrically.* W : strain on $A\ B :: A D : A\ E. :: sin.\ A\ E\ D : sin.\ A\ D\ E.$ Hence, strain on $A\ B = W \frac{sin.\ A\ D\ E}{sin.\ A\ E\ D}$. Again, W : strain on $A\ C :: A\ D : A\ F :: sin.\ A\ F\ D : sin. A D F.$ Hence, strain on $A\ C = W \frac{sin.\ A\ D\ F}{sin.\ A\ F\ D}$. Substituting these values in the formula we have; strain on $A\ B = 1{,}500 \times \frac{sin.\ 60^\circ}{sin.\ 80^\circ} = 1{,}319$. Strain on $A\ C = 1{,}500' \times \frac{sin.\ 40^\circ}{sin.\ 80^\circ} = 979$. Resolving these strains into their vertical and horizontal components, we shall find that the horizontal pressure of one of the beams exactly equals that of the other, whatever be the difference of their inclinations to the vertical, and that the sum of the two vertical components equals the whole weight. The numerical calculation of these components is made as before trigonometrically.

When the span and heights of the struts, and hence their lengths are given, we have more simply, the weight supported at either end $= W \times \frac{\text{non-adjacent segment.}}{\text{Whole span.}}$

That is, the weight supported at $B = W \times \frac{K\ C}{B\ C}$. Weight supported at $C = W \times \frac{B K}{B C}$. Thrust on $A B =$ weight on $B \times \frac{A B}{A K}$. Thrust on $A C =$ weight on $C \times \frac{A C}{A K}$. Horizontal thrust $=$ weight on $B \times \frac{B K}{A K} =$ weight on $C \times \frac{K C}{A K}$.

When the rafters are equally inclined, Fig. 34. Required t, or thrust in the direction of their length. $W : t :: B F :: B D :: 2 B E : B D :: 2 B C : B A$; $t = \frac{1}{2} W \frac{B A}{B C} = \frac{1}{2} W \frac{\text{length}}{\text{rise}} = \frac{1}{2} W \times \text{cosec.} A$. To find h, or horizontal thrust.

Fig. 34.

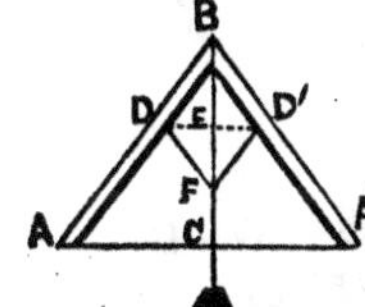

1. By parallelogram of forces. By similar triangles $B E : D E :: B C : A C$, or $\frac{1}{2} W : h ::$ rise : $\frac{1}{2}$ span. Hence $h = \frac{1}{4} W \frac{\text{span}}{\text{rise}} = \frac{1}{2} W \times \text{cotan.} A$.

2. Moment of weight about $A = \frac{1}{2} W \times A C$. Moment of horizontal thrust $= h \times C B$. Hence, $h \times C B = \frac{1}{2} W \times A C$ or $h = \frac{1}{2} W \times \frac{A C}{B C} = \frac{1}{4} W \times \frac{\text{span}}{\text{rise}} = \frac{1}{2} W$ cot. A. If a weight, W, be uniformly distributed, as is that of the rafters themselves, its effect is equal to that of $\frac{1}{2} W$ applied at B, and in that case $t = \frac{1}{4} W \times \frac{\text{length}}{\text{rise}}$, and $h = \frac{1}{8} W \times \frac{\text{span}}{\text{rise}}$. W now means the entire uniformly

distributed load. Otherwise; the horizontal thrust of one beam will be the same as in Fig. 31, that is, the horizontal thrust of $A\,B = \frac{1}{2}$ its wt. $\times \frac{A\,C}{B\,C}$. We might suppose since the horizontal thrust of $A\,B$ is the same as that of $A\,B$, that the horizontal thrusts of the two beams would equal twice the above expression; but such is not the case, for either horizontal force merely corresponds to the reaction of the wall in Fig. 31, consequently, the thrust of both rafters $= \frac{1}{4}$ weight of both $\times \frac{\frac{1}{2}\text{span}}{\text{rise}} = \frac{1}{8}\,W \times \frac{\text{span}}{\text{rise}}$.

The inclination which produces the least horizontal presure is where the angle A or the pitch $= 35° 15'$, or where the height : $\frac{1}{2}$ span : : 1 : $\sqrt{2}$, or where the height is a little more than one-third span. : ; 1 ! 1.414.

Pressure of Water on Lock Gates.

90. A canal lock gate is a frame of two struts. It is analyzed thus, Fig. 35. Let the whole pressure on one gate, of the water acting perpendicularly to it, $= P$. This is resisted by $\frac{1}{2}\,P$, acting at B and $\frac{1}{2}$ at A. Decompose $\frac{1}{2}\,P$, acting at B, into a pressure h parallel to $A\,D$ and another t acting along $B\,A$. Then $\frac{1}{2}\,P\,;h : : E\,F : B\,F$ or $: : B\,D : B\,A$. Hence $h = \frac{1}{2}\,P\,\frac{B\,A}{B\,D} = \frac{1}{2}\,P \div \frac{B\,D}{B\,A} = \frac{1}{2}\,\frac{P}{\text{sin.}\,A} = \frac{P}{2\,\text{sin.}\,A}$. Then again we have, $\frac{1}{2}\,P : t : : E\,F : B\,E : : B\,D : D\,A$. Hence, t

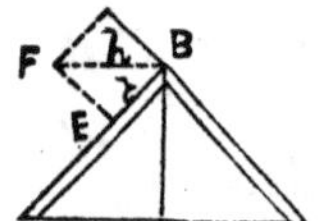

Fig. 35.

$$= \tfrac{1}{2} P \frac{DA}{DB} = \tfrac{1}{2} P. \cot. A = \frac{P}{2 \tan. A}.$$

91. *Braced struts.* The horizontal thrust of a pair of rafters may be resisted by walls, or they may be relieved of it by a tie-beam connecting their feet, as in Fig. 36.

Fig. 36,

When the tie beam is long its middle is supported by a rod or stick, called a "King post." The only strain on it is one-half the weight of the tie-beam. More precisely five-eighths of weight of tie-beam when this is level. (See notes on flexion.)

Influence of change of temperature on iron tie-beams, or tie-rods, of roofs, &c. Let t° = temperature when the frame was constructed, and t°_1 = temperature at a subsequent period. Then $t^\circ - t^\circ_1$= fall of temperature. The shortening (of iron) $= \frac{7}{1000000} \times$ length $\times (t^\circ - t^\circ_1)$. Now an elongation of ,0008 corresponds to a strain of 20,000 pounds per square inch. Then the additional strain caused by cooling will be, $\frac{7}{1000000} \times (t^\circ - t^\circ_1) \times \frac{20000}{.0008} = 175\ (t^\circ - t^\circ_1)$. Thus a change of 100° would cause an extra strain of 17,500 pounds per square inch.

Sometimes the tie-beam is replaced by a "*collar beam*," as indicated by the dotted lines in Fig. 36. In a roof of this kind, the strain on the "*collar-beam*," is greater than that on a tie-beam in proportion to its elevation. If midway up it will strain twice as much, &c. This must be so to satisfy the condition of moments.

12

Fig. 37.

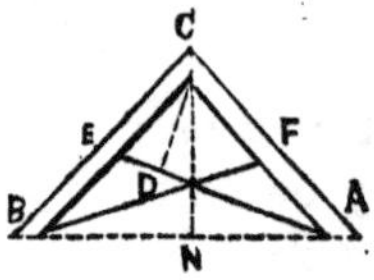

Sometimes the tie beam is replaced by two oblique ties as in Fig. 37. Moment of weight $= \frac{1}{2} W \times \frac{1}{2} A N$. Moment of tie $A E =$ its strain $\times$ $C D$. Hence for equilibrium, $\frac{1}{2} W \times \frac{1}{2} A N =$ strain on tie $\times$ $C D$. Hence strain on this tie must equal $\frac{1}{4} W \frac{A N}{C D} = \frac{1}{8} W \times \frac{A B}{C D}$, or $\frac{1}{2}$ that if there be two ties.

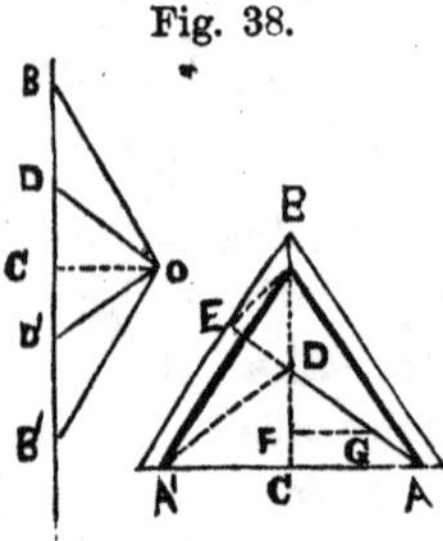
Fig. 38.

In Fig. 38, the tie beam is formed by two pieces, inclined upward and meeting in the middle, and supported by a vertical tie rod. Then this last rod is subjected to a stress from the weight of the rafters, and the load on them. Then the weight W on the vertex $B =$ tension of $B D = W \frac{C D}{B D}$

If $C D = B D$, the tension $= W$. If $C D = 2 B D$, the stress $= 2 W$, &c. Where $C D = 0$, the tension on the rod produced by $W = 0$. The stress on each part of the tie rod $A D$ or $A' D = \frac{1}{2} W \frac{A D}{B D}$. A uniformly distributed load, like that of rafters and roof, produces $\frac{1}{2}$ this stress. To determine the stress on the parts of this frame. Let $G D$ represent the stress on $A D$ in Fig. 38. Decompose it into $G F$ and $D F$. The horizontal component is balanced by that belonging to $A' D$. The vertical component is $D F$. The vertical stress : horizontal stress :: $D F : G F$ or :: $D C : A C$. But the horizontal stress

$= \frac{1}{4} W \times \frac{A\,A'}{B\,C} \times \frac{B\,C}{B\,D} = \frac{1}{2}\,W \frac{A\,C}{B\,D}$. Substituting this into the above proportion, the vertical stress $= \frac{1}{2}\,W \frac{A\,C}{B\,D} \times \frac{C\,D}{A\,C} = \frac{1}{2}\,W \frac{C\,D}{B\,D}$. The other tie $A'\,D$ produces an equal stress on $B\,D$. The total vertical stress on $B\,D$ $= W \frac{C\,D}{B\,D}$. Stress on $A\,D : \frac{1}{2}$ vertical stress at $D :: A\,D : D\,C$, or stress on $A\,D : \frac{1}{2}\,W \frac{D\,C}{B\,D} :: A\,D : D\,C$. Hence stress on $A\,D = \frac{1}{2}\,W \frac{A\,D}{B\,D}$. Stress on $A\,B$ = horizontal stress at $A \times \frac{A\,B}{A\,C} = \frac{1}{2}\,W \times \frac{A\,C}{B\,D} \times \frac{A\,B}{A\,C} = \frac{1}{2}\,W \times \frac{A\,B}{B\,D}$. A uniformly distributed load produces one-half this stress.

Second method. Let W = weight at B, and t = tension of $A\,D$. Draw $B\,E$ perpendicular to $A\,D$ produced. Equating moments, $t \times B\,E = \frac{1}{2}\,W \times A\,C \therefore t = \frac{1}{2}\,W \frac{A\,C}{B\,E} = \frac{1}{2}\,W \times \frac{A\,C}{B\,D \sin. B\,D\,E} = \frac{1}{2}\,W \frac{A\,C}{B\,D \sin. A\,D\,C}$ $= \frac{1}{2}\,W.\,A\,C \div B\,D \frac{A\,C}{A\,D} = \frac{1}{2}\,W \frac{A\,D}{B\,D}$. Tension of $B\,D$ $= 2\,t \frac{C\,D}{A\,D} = 2 \times \frac{1}{2}\,W \times \frac{A\,D}{B\,D} \times \frac{C\,D}{A\,D} = W \frac{C\,D}{B\,D}$.

Third method. Construct a diagram of the forces by drawing from some point *o*, parallels to the bars, intersected by a vertical line. The length $B\,D$ and $B^1\,D^1$ will represent the supporting forces at A and A^1, and their sum equals the total load at B, which equals W. But for equilibrium the load must equal $B\,B^1$. Then the stress on the tie $B\ D$, must be represented by $D\ D^1$. Numerically one-half this stress : $\frac{1}{2}\ W$: : $C\ D : B\ D$. Whence this stress on $B\ D = W \frac{C\ D}{B\ D}$. Stress on $A\ D : \frac{1}{2}\ W$: : $A\ D : B\ D$. Hence stress on $A\ D = \frac{1}{2}\ W \frac{A\ D}{B\ D}$. Stress on $A\ B = \frac{1}{2}\ W \frac{A\ B}{B\ D}$. Horizontal stress : stress of $B\ D$: : $A\ C : 2\ C\ D$. Hence $h = W\ \frac{C\ D}{B\ D} = \frac{A\ C}{2\ C\ D} = \frac{1}{2}\ W\ \frac{A\ C}{B\ D}$.

92. *Trussed Struts.* Roofs of large span are usually subdivided into other triangular frames variously combined. Fig. 39. The struts $E\ F$ and $E^1\ F^1$ are perpendicular to the rafters at their middle points. Hence $A\ E = E\ B$. Let w = weight of one rafter and its load. The horizontal tension h of $E\ E^1 = \frac{1}{8} \times 2\ w\ \frac{\text{span}}{\text{rise}} = \frac{w}{2\ \text{tan.}\ A}$. The parts $A\ E$ and $A^1\ E^1$ besides this have extra tension, t' from the effect of the struts $F\ E$ and $F'\ E'$ which also cause a tension, t'', of the inclined ties $B\ E$ and $B\ E'$. The vertical pressure at $F = \frac{1}{2}\ W$. Its component, R, along $F\ E = \frac{1}{2}\ w$ cos. A. Components ($t' = t''$) of this along $A\ E$ and $B\ E$ (they being equal in length) $= \frac{\frac{1}{2} \times w \text{ cos. } A}{2 \text{ sin. } A} = \frac{w}{4 \text{ tan. } A} = \frac{1}{2}\ h$. This being added to the tension h of $A\ E$ and $A'\ E'$ makes their total horizontal strain $h' = h + \frac{1}{2}\ h = \frac{3}{2}\ h$. Finally

Fig. 39.

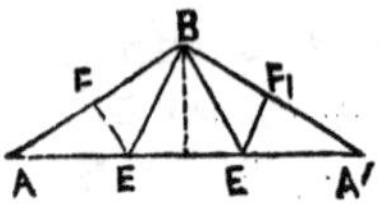

calling $W = 2\,w$, the tension of $E\,E' = \frac{1}{8}\,W\,\frac{\text{span}}{\text{rise}}$. Tension of $A\,E$ and $A'\,E' = \frac{3}{16}\,W\,\frac{\text{span}}{\text{rise}}$. Tension of $B\,E$ and $B\,E' = \frac{1}{2}$ tension of $E\,E' = \frac{1}{16}\,W\,\frac{\text{span}}{\text{rise}}$. These strains could also be obtained by decomposing the strains by graphic construction. Roofs are also sometimes sustained as shown in Fig. 40. The strain on B is $\frac{1}{2}\,W$, on C and $C'\ \frac{1}{4}\,W$, and on D and $D'\ \frac{1}{6}\,W$. The strain on $C\,E$ is $\frac{1}{2}\,W$, and on $B\,F\ \frac{1}{3}\,W$. When the tie-beam is raised (as in Fig. 41), which is usual in iron roofs to give a lighter look, the strain on it is increased in the proportion of former height to new height. The other stresses on it are found similarly to Fig. 38. Draw a diagram by parallels to the bars. Then $I\,F = I'\,F' = \frac{1}{2}\,W$, and the lines $F\,G$ and $F'\,G$ in the diagram, are proportional (on the same scale) to the stress $B\,E$ and $B\,E'$ on the frame. A modification of the preceding is to add the ties shown by the dotted lines, drawn from E and E'. Various arrangements for wide iron roofs for railroad depots, are given in *Morin's "Resistance des Materiaux,"* Pl. VI.

Fig. 40.

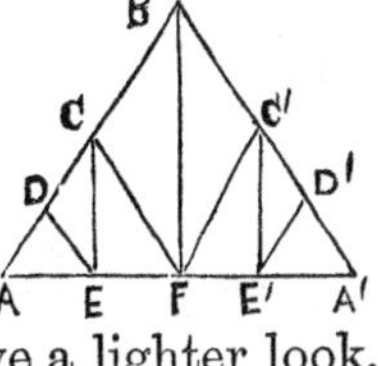

Fig. 41.

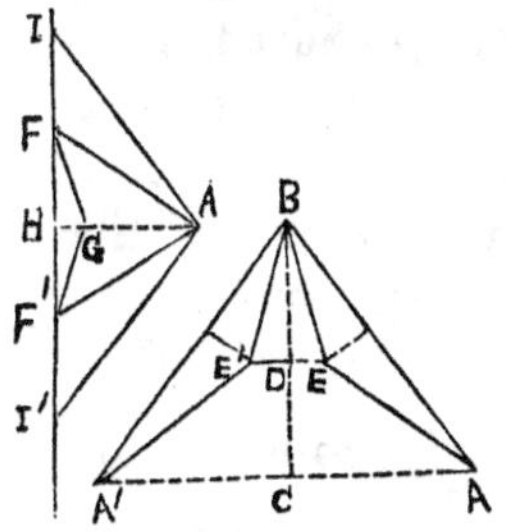

A Pair of Ties.

93. This is just the converse of Art. 88. The stresses are just the same in amount but cause tension instead of compression. Fig. 42. Horizontal strain produced by $W = \frac{1}{4} W \frac{\text{span}}{\text{depression}}$, and the strain on each tie $= \frac{1}{2} W \frac{\text{length}}{\text{depression}}$.

Fig. 42.

Fig. 43 shows a beam $A\ A'$ supported by a strut, $B\ C$, below, and by tie rods $A\ B$ and $A'\ B$ which can be screwed to any tension. With a load W in the middle, the strain on each tie rod, to be in equilibrium with the load, must be equal to $\frac{1}{2} W \frac{\text{length of rod } A\ B}{\text{depression} = B\ C}$.

Fig. 43.

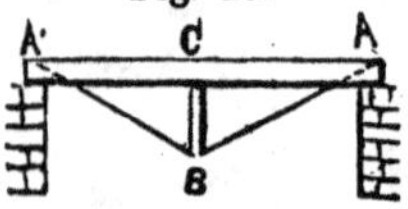

For a passing load, a tension sufficient to resist the action of the load in the centre would exercise the same force after the load had passed to bend the beam in a contrary direction. To make the actual flexure of the beam a minumum let each produce an effect equal to one half this, and the rod should have a tension $= \frac{1}{4} W \times \frac{\text{length of rod}}{\text{depression}}$. The beam will then require to have only one-half the strength to resist flexure which would otherwise be necessary.

A strut and a tie.

94. Such a combination in its simplest form is called a "*bracket.*" This is a frame consisting of two pieces, a

strut and a tie, supporting a load; the two points of support of the strut and the tie being on the same side of the load. In all the forms the upper piece is a tie and the lower a strut. We are to calculate the strain on each. The beams may slope in a contrary direction, or in the same direction. The investigation applies to all.

Fig. 44.

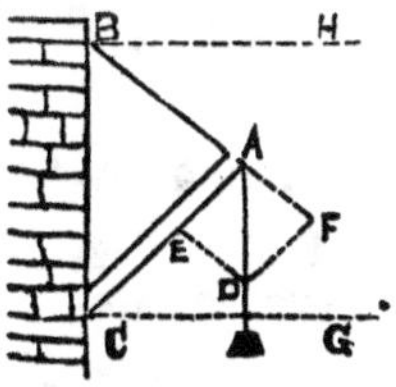

Let $A D$, Fig. 44, represent the units in weight at a scale of one inch for any convenient number of pounds. On $A D$ as a diagonal construct the parallelogram $A F D E$. Then will $A E$ represent the thrust on the strut $A C$, and $A F$ will represent the strain on the tie $A B$. We then have $A F : A D :: A B : B C$. Hence, $A F = W \frac{\sin. A C B}{\sin. B A C} = W \frac{\cos. A C G}{\sin. B A C}$. And $A E : A D :: A C : B C$. Hence, $A E = W \frac{A C}{B C} = W \frac{\sin. A B C}{\sin. B A C} = W \frac{\cos. A B H}{\sin. B A C}$. In words this becomes. The strain on either piece equals $W \times$ cos. of the angle which the other piece makes with the horizon, divided by the sine of the angle which they make with each other.

When one of the beams becomes horizontal the formulas may be simplified. When the tie is horizontal, $B A C = A C G$, and $A B H = 0$, and the above formulas becomes $A F = W \frac{\cos. B A C}{\sin. B A C} = W \times \cot. B A C = W \frac{A B}{B C}$, and $A E = W \frac{\cos. 0^{\circ}}{\sin. B A C} = W \frac{A C}{B C}$. This also will follow directly from the triangle $A\ B\ C$, whose sides will then be parallel to the directions of forces exerted. When the strut is horizontal, $A\ B\ H = B\ A\ C$, and $A\ C\ G = 0$. Then the above formulas become; $A\ F = W \frac{\cos. 0}{\sin. B\ A\ C} = \frac{W}{\sin. B\ A\ C} = W \frac{A\ B}{B\ C}$, and $A\ E = W \frac{\cos. B\ A\ C}{\sin. B\ A\ C} = W \frac{A\ C}{B\ C}$. In this form the strain is greater on the tie than on the strut. The expressions for $A\ F$ and $A\ E$ being interchanged. The preceding form is therefore preferable, because any defect in the wood lessens its power of extension more than compression.

95. *Modification of brackets.* A beam fixed at one end and supported between its ends. Fig. 45. The weight acts in the line $D\ G$. The strut acts in the line $C\ A$, meeting $D\ G$ produced in E. The force at B must act in the line $B\ E$, since three forces in equilibrium must meet in a common point.

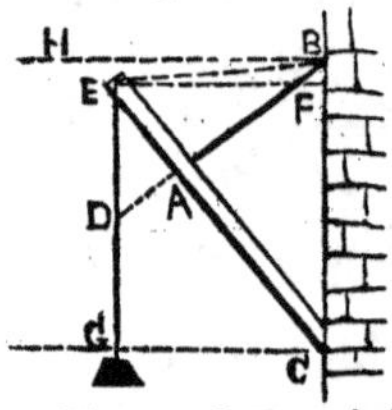

Fig. 45.

Since the weight acts parallel to $B\ C$ the three forces in equilibrium are parallel to the three sides of the triangle $B\ C\ E$, and proportional to them. $B\ C$ will represent the units in the weight, $C\ E$ the units of pressure on the strut, and $B\ E$ the force at B in the direction $B\ E$. Pressure on the strut at C in the direction $A\ C = W \frac{C\ E}{C\ B}$. The force at B in the direction $B\ E = W \frac{B\ E}{C\ B}$. *Required; the hor-*

izontal and vertical components of these strains. The force $C\,E$ may be resolved into $C\,F$ and $F\,E$ and the force $B\,E$ into $B\,F$ and $F\,E$; hence we see that the horizontal thrust at C and B are equal but in opposite directions. The vertical forces acting at C and B are as $C\,F$ and $B\,F$; $C\,F$ acting downwards and $B\,F$ upwards. Their difference, $C\,B$, represents the weight. There are many forms of brackets, depending on the directions of the strut and tie, and their angle with each other; but the above formulas apply to all, by making the proper modifications for the different conditions.

Fig. 46.

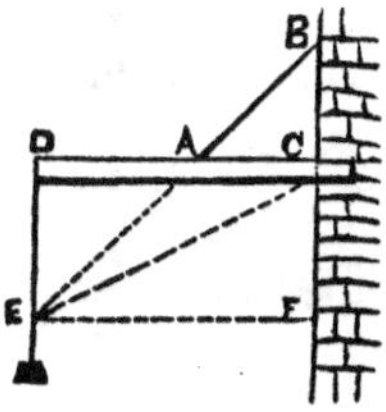

When the beam is horizontal as in Fig. 46, the strains are as the sides of the triangle $B\,C\,E$.

The pressure at $B = W\dfrac{B\,E}{B\,C} = W\dfrac{B\,E}{B\,A} \times \dfrac{B\,A}{B\,C}$, but $\dfrac{B\,E}{B\,A} = \dfrac{E\,F}{C\,A} = \dfrac{C\,D}{C\,A}$. Then the pressure at $B = W \dfrac{C\,D}{C\,A} \times \dfrac{B\,A}{B\,C}$.

Trigonometrically we have $\dfrac{B\,A}{B\,C} = \text{cosec.}\,B\,A\,C$.

Then the pressure at

$$B = W\frac{C\,D}{C\,A} \times \text{cosec.}\,B\,A\,C = W\frac{C\,D}{C\,A\,\text{sin.}\,BAC}.$$

To find the horizontal pressure at C, resolve the pressure at B into BC and $C\,A$ and we get, horizontal pressure = pressure at

$$B \times \frac{C\,A}{A\,B} = W\frac{C\,D}{C\,A} \times \frac{B\,A}{B\,C} \times \frac{C\,A}{B\,A} = W\frac{C\,D}{B\,C}.$$

Otherwise. The weight W, acting at D, is equivalent to $W\frac{C\,D}{C\,A}$ acting at A. Let $B\,C$ equal this. Then the other strains are proportional to BA and $C\,A$, and we have; strain on $B\,A = W\frac{C\,D}{C\,A} \times \frac{B\,A}{B\,C}$. Strain on $A\,C = W\frac{C\,D}{C\,A} \times \frac{C\,A}{B\,C} = W\frac{C\,D}{B\,C}$. When the tie is horizontal, the same formulas apply, interchanging the letters B and C. Stress on $C\,A = W\frac{B\,D}{B\,A} \times \frac{C\,A}{B\,C}$ and horizontal stress at $B = W\frac{B\,D}{B\,C}$.

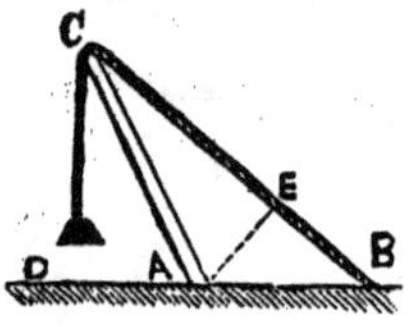

Fig. 47.

Another case of a strut and a tie is shown in Fig. 47. The triangle $A\ C\ E$ has its sides parallel to the direction of the stresses, and hence we find the piece $A\ C$ is compressed with a force equal to $W\frac{\cos.\ C\ B\ D}{\sin.\ (C\,A\,D - C\,B\,D)}$.

The tension of the rope $B\,C = W\frac{\cos.\ C\ A\ D}{\sin.(C\ A\ D - CB\,D.)}$ The horizontal strain tending to make the piece $A\ C$ slide at $A = W\frac{\cos.\ C\ B\ D.\ \cos.\ C\ A\ D}{\sin.(C\ A\ D\ -\ C\ B\ D)}$. If the weight was not merely suspended at C, but supported by a cord passing through one or more pulleys (as in the case of a crane), we should have to consider instead of W, the resultant of W and of the tension of the cord.

Two struts and a straining beam.

96. With loads W and W' on C and C' the stresses are analagous to Fig. 34. Consider the upward reaction at A and A' which are each equal to W. Then the triangle $A\ B\ C$ gives stress on

Fig. 48.

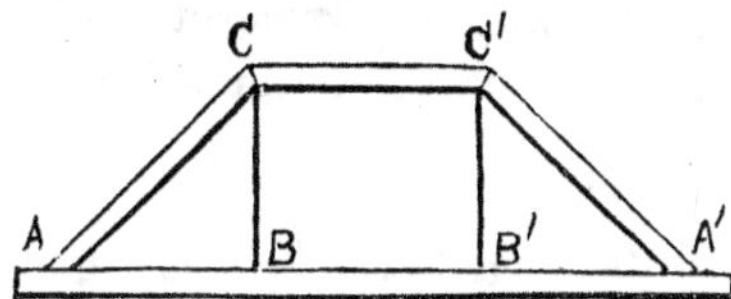

$A\ C = W \dfrac{A\ C}{B\ C}$, horizontal stress $= W \dfrac{A\ B}{B\ C}$. All the stresses are the same as if the struts, $A\ C$ and $A'\ C'$, were produced upward to meet, and the whole load $(W + W')$ placed at that point, as in Fig. 34.

Suppose the load on $A\ A'$ to be uniform, and supported by rods $B\ C$ and $B'\ C'$, $A\ A'$ not to be continuous but to be divided at B and B'. This corresponds to the weight on $C\ C'$ in Fig. 48. If the beam $A\ A'$ be continuous and level, then $\frac{11}{30}\ W$ is on $B\ C$ and $B'\ C'$, and $\frac{4}{30}$ on A and A'. This is safest to take, being greatest, though it is not generally done.

In either case calling W' the load on $B\ C$ or $B'\ C'$, then the horizontal thrust on $C\ C'$ and the pull on $A\ A' = W' \dfrac{A\ B}{A\ C} = W' \frac{1}{3} \dfrac{A\ A'}{B\ C} = \frac{1}{3} W' \dfrac{\text{span}}{\text{rise}} = \frac{1}{9}\ W \dfrac{\text{span}}{\text{rise}}$. The thrust on $C\,A$ and $C'\,A' = W' \dfrac{A\,C}{B\,C}$. If a similar load were on top, the stresses would be the same.

When a bridge of this form is reversed the stresses remain the same except that the former stresses of compression have become extension, and *vice versa.* This arrangement may be extended to any number of panels. It is preferable for materials like wood and wrought iron, because the shortest pieces are exposed to compression.

Multiple Frames.

97. Suppose a uniform load W on a beam $A\,A'$, Fig. 49. Each post or vertical tie supports $\frac{1}{4}\,W = W'$. The struts $E\,B$ and $E\,B'$ resist a stress $= \frac{1}{2}\,W' \dfrac{B\,E}{C\,E}$, as found from Fig. 32. They produce a horizontal strain at E and on $B\,B' = \frac{1}{2}\,W' \dfrac{B\,C}{C\,E}$. The weight on D or $D' = W' + \frac{1}{2}\,W' = \frac{3}{2}\,W'$ ($\frac{1}{2}\,W'$ being transferred from E.) Consequently $D\,A$ and $D'\,A'$ have a stress $= \frac{3}{2}\,W' \dfrac{A\,D}{B\,D}$. They produce a horizontal strain on $D\,D'$ and $A\,A' = \frac{3}{2}\,W' \dfrac{A\,B}{D\,B}$. The horizontal strain $B\,B' =$ the sum of the two horizontal strains $= (\frac{1}{2}\,W' + \frac{3}{2}\,W')$ $\frac{1}{4}\,\dfrac{\text{span}}{\text{rise}} = \frac{1}{2}\,W \dfrac{\frac{1}{4}\,\text{span}}{\text{rise}} = \frac{1}{8}\,W \dfrac{\text{span}}{\text{rise}}$.

Fig. 49.

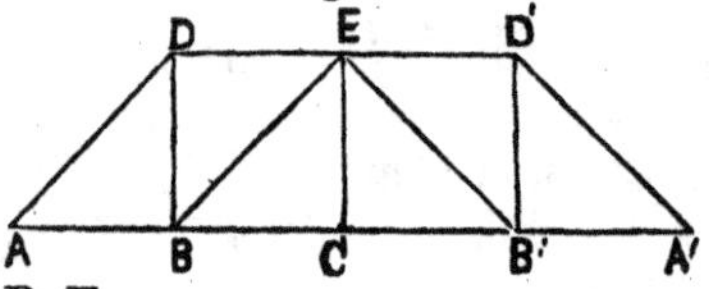

So proceed for any number of panels. It will be found that the strains on the posts and on the struts, increase in a direct ratio to the distance from the centre. Their strength and size should therefore be increased in the same ratio. The strains on the top and bottom beams increase from the ends to the middle, but not in a direct ratio. The increase is most rapid as you proceed from each end and becomes less rapid on approaching the centre or middle. It is analogous to that of a solid beam, in which latter case the relative increase is indicated graphically by a parabolic curve (see Fig. 21.)

The usual formula for the horizontal stress on a frame, caused by a uniform load $\left(\frac{1}{8} W \frac{\text{span}}{\text{rise}}\right)$ supposes the weight to be uniformly applied at the ends of the struts, as well as uniformly over the road way. This is the case in frames which have an even number of panels; but is not so with those of an uneven number. For example with three panels, the horizontal stress $= \frac{1}{9} W \frac{\text{span}}{\text{rise}}$; for five panels $= \frac{1}{8\frac{1}{3}} W \frac{\text{span}}{\text{rise}}$, for seven panels $= \frac{1}{8\frac{1}{6}} W \frac{\text{span}}{\text{rise}}$; and generally for n panels (n being uneven) $= \left(\frac{1}{8} - \frac{1}{8 n^2}\right) W \frac{\text{span}}{\text{rise}}$.

In the preceding forms the ties were vertical and the struts inclined. In Fig. 50, both the ties and struts are inclined. The stresses, however, follow similar laws. With a uniform load W, such as its own weight, the vertical strain increases uniformly from the middle, where it equals zero, towards the end where it equals ½ W. At x' from end, or x'' from middle, it equals $W \cdot \frac{x''}{\text{span}}$. The strain on any diagonal whose middle is x'' from the middle of the bridge

Fig. 50,

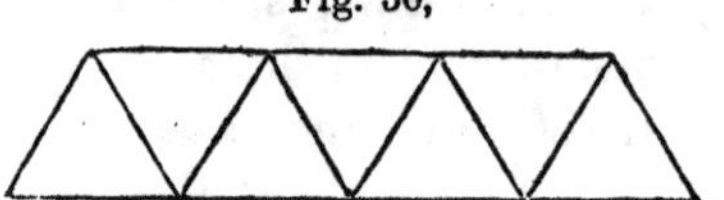

$$= W \frac{x''}{\text{span}} \times \frac{\text{length of diagonal}}{\text{depth of panel}}.$$

The horizontal strain at the centre $= \frac{1}{8} W \frac{\text{span}}{\text{rise}}$. At any point x'' from middle or x' from end, it

$$= \frac{W}{2} \times \frac{x' (\text{span} - x'')}{\text{span} \times \text{depth}}.$$

diminishing from the centre to the ends in ratio before shown by a parabolic curve.

When the loads are applied along top or bottom, or along both. The distance x' and x'', in the preceding formulas, are measured to the tops of the diagonals when the load is attached to the bottom of the beam, to the lower ends of it if it be on top, and to their middle if the load be equally on the top and bottom, as its own weight. Bridges of this form are called "Triangular Girders," or "Half Lattice," or "Warren's," or "Neville's." If the number of the oblique pieces be doubled, then each sustains half the above strains. This is a lattice bridge. Any number of such pieces may be combined.

98. *Best angle for the diagonals in a triangular girder, when all make the same angle with the horizon.* Their total length is less in proportion to the smallness of their angles with the horizon. The strain on them is greater in the same ratio, and this increases their necessary cross-section. Then, there will be some one angle which will require the least amount of material and, consequently weight and cost.

Let the angle $B\,A\,C = a$, in Fig. 51, be this angle. Call the weight $B\,C = 1$. Then the length $A\,B = B\,C.$ cosec. a = cosec. a. The strain, and consequently the cross-section, is also as *cosec. a.* Combining these, we have, the material in one varies as cosec.^{2} a. The reach, or horizontal pressure, of each = cot. a. Then their number (calling length of girder $= l$) is $\frac{l}{\text{cot.}\,a}$ Then the total material is as

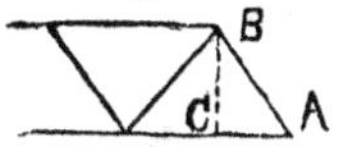

Fig. 51.

$$\text{cosec.}^2 a \times \frac{l}{\text{cot. } a} = \frac{l}{\text{sin.}^2 a} \div \frac{\text{cos. } a}{\text{sin. } a} = \frac{l}{\text{sin. } a \text{ cos. } a}.$$ This is to be a minumum, or sin. $a \times$ cos. a, a maximum. It is so when $a = 45°$.

Other practical considerations render it desirable to make the angle greater. It is usually between 45° and 60°.

Best angle when the diagonals are alternately vertical and oblique.

In Fig. 52. Let $B A C = a$. Calling the depth of the bridge, or the length of the vertical beam, $B C = 1$, the length of the oblique beam, $A B =$ cosec. a. The stress on the oblique beam (calling that on the vertical $= 1$), is also as cosec. a. Their weight, &c., is, consequently, as cosec.^{2} a. The number of each kind of braces (calling the length of bridge $= l) = \frac{l}{\text{cot. } a}$. Then the weight of all the vertical braces are as $\frac{l}{\text{cot. } a}$, and those of the oblique braces as cosec. $a \times \frac{l}{\text{cot. } a}$. Hence the weight of all are as $\frac{l}{\text{cot. } a}$ $(1 + \text{cosec.}^2\ a]$

Fig. 52.

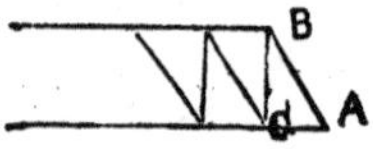

This is to be a minimum. It is so when $a = 35° 16'$. We have been assuming that the weight, length, cross-section, material, &c., of each unit of length of the vertical brace was capable of conveying the same strength as a unit of the oblique brace. If not, let the strength of the former be to that of the latter :: $n : 1$. Then in the preceding expression the weights, &c., will not be as $1 :$ cosec.^{2} a, but as $n :$ cosec.^{2} a. Consequently the expression to be minimum is $\frac{l}{\text{cot. } a}(n + \text{cosec.}^2 a)$. This is a minimum when cot. $a = \sqrt{(1 + n.)}$ When $n = 1$, as in our first hypothesis, cot. $a = \sqrt{2} = 1.414$, and hence, $a = 35° 16'$ as before. When $n = \frac{1}{2}$, as might be for wood and wrought iron where the oblique braces are compressed, then $a = 50° 46'$. When $n = 2$, as might be for cast iron, $a = 60°$.

Compound frames.

100. In this combination, shown in Fig. 53, the stress on each pair of struts may be decomposed, as in Fig. 33, or they may be found by one of the methods of Art. 89. If the frame be inverted the stresses are the same, but the parts before compressed are now extended, and *vice versa.*

Fig. 53.

In the frame, shown in Fig. 54, composed of four distinct trusses, the thrust is found as before. The strain on the tie beam A A', is the sum of that due to each set of struts. The strain on the straining beam, B B', increases from the ends to the middle, the outer portions receiving the stress of the outer pair of struts, the next portion the additional stress of the next pair, and so on. Latrobe's Bridge is on this plan.

Fig. 54.

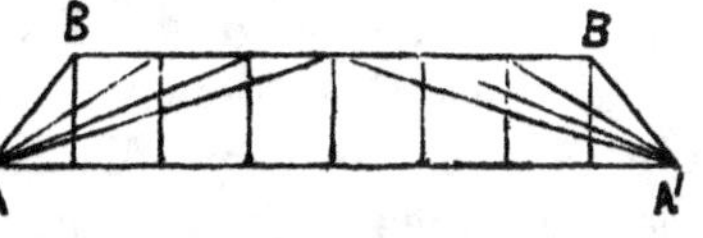

In such a frame as shown in Fig. 55 (sometimes used for a turning bridge) the platform A A', being supported by oblique iron rods, the strain or pull on each rod = load at lower end × length ÷ by the height of its upper end above the lower. The stress, or thrust on the lower beam, A A',

Fig. 55,

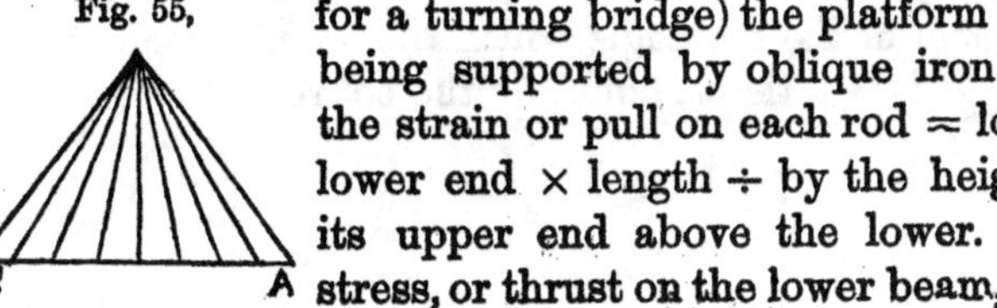

increases from the ends to the middle. Between the two outside tie rods it equals the load at $A \times$ distance from centre $\div$ height of tops of outer rods. So for each remaining rod.

Polygonal Frames.

101. Such frames, even when of only four sides, are liable to change their angles from the action of a slight force. Let the force W act in the direction $E\ B$, Fig. 56, and the corresponding reaction act in $A\ F$. Let the frame be stiffened by the brace $C\ D$ across one of its angles. Required the strain on the brace. The force W tends to rack the frame, that is, tends to turn it around A. Its moment therefore $= W \times A\ M$. Let $P =$ force with which the brace resists, and let $A\ N$ be the perpendicular to the brace. Then the moment of brace $= P \times A\ N$. Hence, for equilibrium,

Fig. 56.

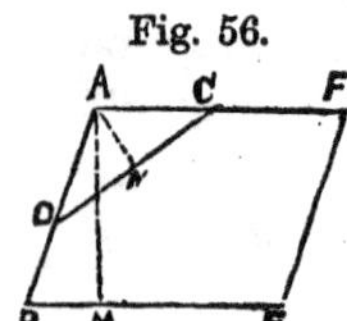

$$P \times A\ N = W \times A\ M \text{ or } P = W \frac{A\ M}{A\ N}.$$ From this we see that the nearer the brace is to A, the more it is strained. Every frame having more than three sides, should be braced so as to form triangles. In a frame like a gate, where the brace is a strut, it should be placed in that diagonal of the rectangle which the weight tends to shorten. If it is a tie it should be the converse.

Fig. 57.

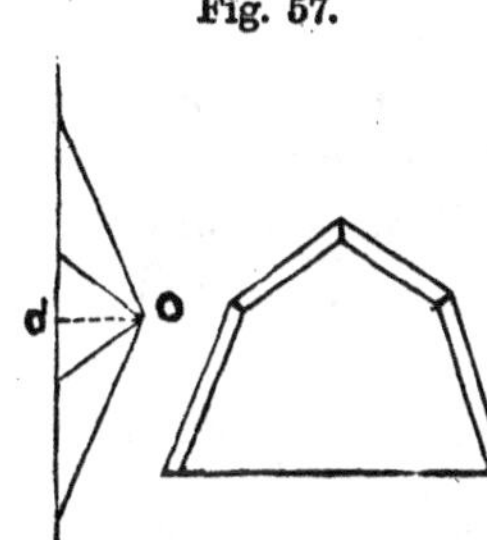

The form in Fig. 57 is called a "curbed," or "Mansard," or "hipped" roof. Practically to find the best angles for such a roof. Prepare four sticks, proportional to the intended length of the rafters, and fastened by pins; attach their extremities to a vertical board, so that when inverted they may hang freely. The position which they then assume gives their proper form in a roof. If the rafters are to be loaded unequally, weights, proportional to their respective loads, should be suspended from the centres of gravity of the pieces.

In such a frame, in which the lower pieces made an angle of 3° with the vertical, and the upper made an angle of from 45° to 63° with the vertical, the horizontal thrust of the former was from .197 to .227 of weight, or about $\frac{1}{5}$, the weight being uniformly distributed along the length of the two upper pieces.

Geometrically. Where a polygonal frame is in equilibrium the horizontal thrust is the same at each angle. Let frame be loaded at each angle. The strains on each beam, in the direction of its length, are proportional to lines drawn from any point, *O*, parallel to those directions, and limited by a vertical line, as shown in Fig. 57. The portions of the vertical line cut off by these diverging lines, are proportional to the weights suspended from the joints of the beams parallel to those lines. *O O′* is proportional to the horizontal stress.

Trigonometrically. The strain on any beam in the direction of its length = horizontal strain × secant of its angle with the horizon. The weight suspended from any angle = horizontal strain × difference of the tangents of the angles which the two beams, meeting at that point, make with the horizon.

NOTE.—The general principles of inflexible frames are these : 1st. Strength and economy are the two objects sought. 2d. Give little excess of strength to the parts which themselves add to the load to be supported. 3d. Endeavor so to arrange the beams that the strain shall be transmitted as nearly as possible through their axes. 4th. Avoid as much as possible transverse strains. 5th. Load obtuse angles lightly. 6th. Divide polygonal frames into triangles by means of braces. 7th. Distinguish carefully between struts and ties and construct them accordingly. 8th. Give all the parts of the frame equal relative strength, since the strength of the frame can never exceed the strength of the weakest part, and partial strength produces general weakness.

CURVED FRAMES.

Convex upwards, or Linear arches.

102. Conceive a polygonal frame to have the number of its sides indefinitely increased and their lengths indefinitely diminished. It then becomes a linear arch. A diagram of forces may by constructed for it, as for a polygonal frame. The various lines drawn from some point O, as in Fig. 57, will then represent the thrusts at those parts of the arch to whose tangents they are respectively parallel. If i be the angle of one of these tangents the thrust there equals the horizontal thrust at the top, multiplied by sec. i.

If the arch sustains a vertical load, uniformly distributed along its length, its form when in equilibrium would be a "catenary." If the arch sustains a vertical load distributed horizontally, its curve when in equilibrium would be a parabola. If the arch sustains a uniform normal pressure, like that of a fluid, its form would be circular. If the arch sustains a pressure not uniform, but symmetrical on opposite sides, its form would be elliptical and the ratio of the axes should be the square root of the two pressures. Thus, for a tunnel arch very deep under ground we should use an ellipse in which the horizontal semi-axis, divided by the vertical semi-axis = the square root of the horizontal pressure of earth, divided by the square of the vertical pressure of earth.

An arch sustaining a pressure every where proportional to the depth of the point in question below a horizontal plain (such is the pressure of standing water), should have a curve such that the radius of curvature at the point is inversely proportional to the depth. This curve somewhat resembles a cycloid. It has been called a "*hydrostatic arch.*" An approximate form for an hydrostatic arch is a semi-ellipse of the same height, and having its greatest and least radii of curvature, in the ratio of the greatest and least depth of load. Let x_1 = depth of load at the crown of such an arch. Let x_{11} = depth of load at the spring. Then approximately its half span = $\frac{19}{20}(x_{11} - x_1)\frac{\sqrt{x_{11}}}{\sqrt{x_1}}$. An arch sustaining a pressure of earth at any depth should have a form named the "Geostatic arch." Its equation is very complicated. (For a full investigation of these last two curves, see Rankin's Applied Mechanics, pp. 190–208.)

Experiments on wooden arches formed of bent planks, so that they could approximately be considered linear arches, gave the following results, the arches being semicircular: When the weight is uniformly distributed along the circumference, the horizontal thrust = about one-sixth weight. This rule also gives their thrust by their own weight. When the weight is uniformly distributed horizontally, the thrust = about two-ninths weight. When the weight is all at the summit of the arch the horizontal thrust = about one-third weight. When the weight rests vertically above a point one-fourth the diameter from the spring, the horizontal thrust is about two-sevenths weight. If the arch be depressed or elevated the thrust varies as the ratio, one-half span : rise. Frames of straight pieces are preferable to curves for a given amount of material. Structures framed of straight pieces resist flexure twice as well as those of curved pieces.

Cast iron, timber and similar arches, are usually arches only in form. Such an arch may be regarded as composed of two parts resting against each other at the crown.

Let G, Fig. 58, be the centre of gravity of the semi-arch. Its weight acts through the verticle drawn through G. From the middle point of the crown, H, draw a horizontal line meeting the vertical through G in F. From F draw a line $F\,I$ to the spring at its middle point, and draw $I\,M$ parallel to $F\,H$. Set off $F\,G$ = weight of semi-arch, and complete the parallelogram $F\,L\,G\,K$. Then $F\,K$ will re-

Fig. 58.

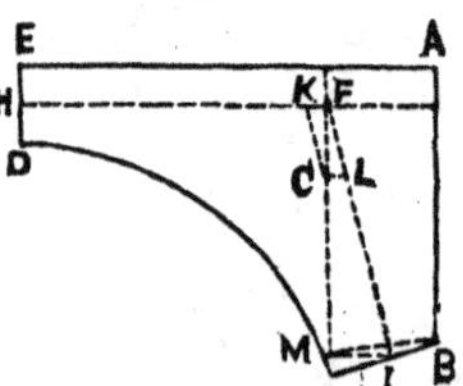

present the horizontal thrust, and $F\,L$ its pressure on the abutment. By similar triangles we have $F\,K = \frac{I\,M \times F\,G}{F\,M}$

Hence, horizontal thrust = horizontal distance from the middle of the spring to the vertical through the centre of gravity of the semi-arch, ×. weight of semi-arch ÷ vertical distance from the centre of the spring to the centre of the crown.

Approximately, calling centre of gravity mid-way between spring and crown, the horizontal thrust = $\frac{1}{8}$ weight of whole arch × $\frac{\text{span}}{\text{rise}}$. This errs on the safe side. The oblique thrust at the abutment = $\sqrt{}$(horizontal thrust^2 + weight $\frac{1}{2}$ arch^2.)

103. *Concave upwards or Catenaries.* Let x = depth of curve, $y = \frac{1}{2}$ span, t = horizontal tension at the lowest point, T = tension at point of suspension, a = angle which the tangent at the point of suspension, makes with horizon, and l = length of curve. Let p = the length of a similar chain whose weight = tension at the lowest point.

Given x and y to find a°. A long investigation gives $\frac{\text{log. tan. } (45^\circ + \frac{1}{2} a)}{\text{sec. } a \times \text{ver. sin. } a} = \frac{y}{2.3\,x}$. The method of finding the value of a from this formula, is by trying different values for it and continually approximating until one is found which satisfies the equation. Care must be taken to subtract ten from the log. tan. of the numerator, in order to make it of the same radius as the denominator, that is, to make the equation homogenous. We also find from the same investigation the formulas:

$$p = \frac{x}{\text{sec. } a - 1}, \; T = \frac{x}{\text{ver. sin. } a}, \; l = \frac{2\,x \text{ sin. } a}{\text{ver. sin } a}.$$

The weight and consequently the whole length of any portion of a uniform chain will be proportional to the tangent of the inclination of the catenary at the upper end of this portion. This is the characteristic property of the catenary. The strain exerted tangentially, at any point, is proportional to the secant of the inclination at that point. Practically, the catenary does not differ sensibly from the parabola as applied to suspension bridges, since in them the platform of the bridge and the load upon it are uniformly distributed along a horizontal line, and not along the chain. The curve thus approximates nearer a parabola than a catenary, and would be an exact parabola if its own weight as well as that of the platform were distributed horizontally.

The Catenary as a Parabola.

Fig. 59.

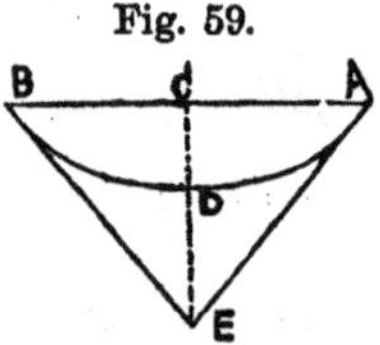

Let the span $A\,B$ of Fig. 59, equal $2\,y$, the deflection $= x$, $t =$ horizontal tension at D, and $w =$ weight of catenary per unit of span. Required; horizontal strain at D. Moment of horizontal strain $= t\,x$. Moment of chain itself $= w\,y \times \frac{y}{2} = \frac{w\,y}{2} \times y$;

Hence, for equilibrium $t\,x = \frac{w\,y}{2} \times y$, $t = \frac{w\,y \times y}{2\,x} = \frac{1}{2}$wt.

of chain $\times \frac{1}{4} \frac{\text{span}}{\text{rise}} = \frac{1}{8}$ wt. of chain $\times \frac{\text{span}}{\text{rise}}$. This investigation assumes the weight to be uniformly distributed horizontally, in which case the curve will be a parabola, having for its equation $y^2 = \frac{2\,t}{w\,x}$. Hence in such a parabola, $\frac{2\,t}{w}$ = parameter.

2d. To find the tension at any other point. It is the resultant of the horizontal tension and of the weight between the required point and vertex. This resultant is the hypothenuse of a right angled triangle, and is numerically equal to the square root of the sum of the squares of the two forces. Let y_1 = ordinate of any point as D_1; then the weight of the portion $D\,D_1 = w\,y_1$, making t = one side of a right angled triangle, and the weight of DD_1 as the other side, then t_1 or hypothenuse $= \sqrt{[t^2 + (wy_1)^2]}$. The tension at the point of suspension, represented by

$$T_{,} = \sqrt{t^2 + w\,y^2} = \sqrt{\left(\frac{w^2\,y^4}{4\,x^2} + y^2\,w^2\right)} = w\,y\sqrt{\left(\frac{y^2}{4\,x^2} + 1.\right)}$$

3d. To find the angle which the tangent to this curve makes with the horizon, i. e., $C\,A\,E$. When the curve is a parabola we have, tan. $C\,A\,E = \frac{C\,E}{C\,A} = \frac{2\,C\,D}{C\,A} = \frac{2\,x}{y}$ $= 2\,\frac{\text{deflection}}{\frac{1}{2}\text{ span}}$. An approximate formula for the length of curve is, $2\,y + \frac{4\,x^2}{3\,y}$, or span $+ \frac{8 \text{ sq. of depth}}{3 \text{ span.}}$

4.th Depression caused by elongation, produced by stretching, as by heat. Depression in middle equals elongation $\times \frac{3}{16} \times \frac{\text{span}}{\text{ver. sin.}}$.

5th. When produced by a weight in the centre; depression $= \frac{\text{weight in centre} \times \text{ver. sin.}}{2\,(\text{wt. of chain} + \text{load}) + \frac{1}{4}\,\text{wt. in centre.}}$

Fig. 60.

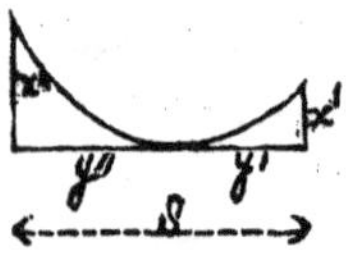

6th. Chain suspended at unequal heights, considered as a parabola. Fig. 60.

$$y_1 = S \times \frac{\sqrt{x_1}}{\sqrt{x_1} + \sqrt{x_{11}}}$$

$$y_{11} = S \times \frac{\sqrt{x_{11}}}{\sqrt{x_1} + \sqrt{x_{11}}}.$$

$$\text{Horizontal tension} = \frac{W S^2}{2 x_1 + 2 x_{11} + 4 \sqrt{x_1 x_{11}}}.$$

$$\text{Length} = S + \tfrac{2}{3}\left(\frac{x_1^2}{y_1} + \frac{x_{11}^2}{y_{11}}\right)$$

Note on Models.

104. *The relations between the strength of frames and small models of them.* No model can have all of its parts enlarged in the same ratio and be as strong, comparatively, as before. There are three different cases.

1. *Case of extension.* Suppose the structure ten times the size of the model. The cross-section, and strength of a piece in the structure is increased one hundred times that in the model, but its load has been increased one thousand times. Suppose the model could have supported

twice the original load upon it, the full sized structure can bear only one-fifth the increased load. A good example of this is a suspension bridge. It will, say, with 1,000′ span, bear a heavy load, but increase its span eight or ten times and all its parts proportionally, and it will fall by its own weight. Large spiders spin threads much larger in proportion to their thickness than small ones. Suppose one nine times as thick as the other, his weight would be (9^3) times that of the smaller. His thread would be twenty-seven times as thick.

2. *Case of compression as a prop or support.* A model being increased n times as before, its strength is increased n^2 times, whilst its load is increased n^3 times. Nature affords an example of this kind also. Were every part of a man enlarged proportionally until his body equalled that of the elephant, his legs would break under his own weight. Hence the great proportional size of the elephant's legs.

3. *Case of cross-section.* For this we have the general formula $W = S\frac{b\,h^2}{l}$, in which b, l, and h, are the dimensions of the beam and S the coefficient found by experiment. Required, what will be the effect of increasing all the dimensions n times, $W = S\frac{n\,b\,(n\,h)^2}{n\,l} = S\frac{n^2\,b\,h^2}{l}$. This shows that the strength has been increased n^2 times, but the load has increased n^3 times, and the beam is therefore only $\frac{1}{n}$ as strong as before. Therefore for cross strain in enlarging a model n times, the breadth must

again be increased n times, that is, n^2 times as great as originally, if the depth be increased n times only. The depth must be made n t n times as great as originally, if the breadth be increased n times. Although increasing the dimensions of the beam n times, makes it n^2 times stronger to resist breaking, yet it becomes only n times stronger to resist bending, since we have,

$$\delta = \frac{W}{c} \times \frac{(n\,l)^3}{n\,b\,(n\,h)^3} = \frac{1}{n} \times \frac{W}{c} \times \frac{l^3}{b\,h^3},$$

that is, $\frac{1}{n}$ of the former deflection, the load being supposed to remain the same. Hence the deflection of beams from their own weight, or from a uniform load proportionally increased (that is n^3 times) will be $\frac{n^3}{n}$ or as the square of the ratio of increase. To preserve the same stiffness against deflection of beams, we see by the formula that the depth must increase in the same ratio as the length, if the breadth remains constant, or if the depth remain the same, the breadth must be increased as the cube of the length. For example; a beam whose length has been increased ten times, must, in order to retain the same stiffness as before, be ten times as deep, or one thousand times as broad.

Calculation of the strength of structures from that of their models. Let p = weight that the model will bear at its centre, w = weight of model, r = ratio of the scale of the model to that of the structure, P = load the structure must be able to bear at the centre. Then it is evident that $r^3\ w$ = weight of structure. Resistance of structure : resistance model : : cross-section of structure : cross-section of model, or : : $r^2 : 1$.

A weight uniformly distributed = ½ same weight at the centre. Hence $p + \frac{1}{2} w$ = resistance of model, and since the resistance of the structure = r^2 times that of model we have $P = r^2 (p + \frac{1}{2} w) - \frac{1}{2} r^3 w = r^2 p - \frac{r^3 w - r^2 w}{2}$

$$= r^2 p - r^2 (r - 1) \frac{w}{2} = r^2 \left[p - (r - 1) \frac{w}{2} \right].$$

Conversely.—P given and p required. From preceding formula we have, $p = \frac{P}{r^2} + (r - 1) \frac{w}{2}$.

BLOCK WORK OR STRUCTURES WITH WIDE JOINTS.

Single pieces, or cemented masses.

105. *General principles of stability.* A solid resists overturning by its own weight. This action may be considered as concentrated at, and acting in a line passing through, its centre of gravity. The "*moment of stability*" of a body = its weight, × horizontal distance from this vertical line to the edge about which the body would turn if overthrown.

In Fig. 61, the moment of the wall $A\ B\ C\ D$ = weight × $A\ H$. In all cases where the contrary is not mentioned we consider a portion of wall one foot long. The moment of any force, as P, tending to overthrow it = P × length of the line drawn from the front edge of the wall, perpendicular to the direction of the force or = $P \times A\ F$.

Fig. 61.

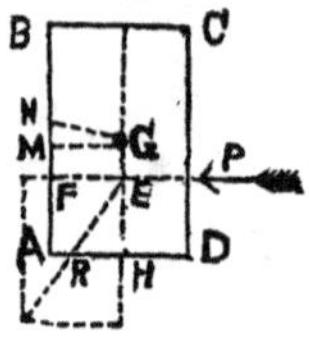

There are two ways of determining whether a wall will stand or fall under a given pressure.

1. *Graphically.* Construct a parallelogram of forces at the point where the direction of the force meets the vertical through the centre of gravity. Draw its diagonal. If this passes within the front edge the wall will stand; if not, it will fall.

2. *Numerically.* Calculate the moments of the wall, and of the force. If the former exceed the latter, the wall will stand, and *vice versa.* If the force acts horizon-

tally, its leverage is the height of the point of application above the edge.

If force acts obliquely, Fig. 62, the leverage is obtained thus: *Numerically*. It will be in equilibrium when $P \times A\,F = W \times A\,H$. Required to find $A\,F$. Let $a =$ angle which the direction of the force makes with the horizon. Then $A\,F = A\,V \times \cos.\ a = (C\,D - C\,Z) \cos.\ a = C\,D \cos.\ a - C\,Z \cos.\ a$. $C\,Z = V\,Z \times \tan.\ a = V\,Z \times \frac{\sin.\ a}{\cos.\ a}$. Hence, $A\,F = C\,D \times \cos.\ a - V\,Z \times \frac{\sin.\ a}{\cos.\ a} \times \cos.\ a = C\,D \times \cos.\ a - V\,Z \times \sin. a$. That is, the perpendicular from the point of overturning to the direction of the force = height of the point of application of the force × cos. a — the distance from the point of overturning to the vertical through point of application of the force, × sin. a.

Fig. 62.

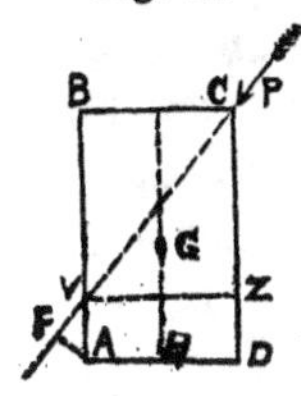

The "*moment of stability*" of a body, $= W \times H\,A$. The "*dynamical stability*" of a body $= W \times$ height, through which it is necessary to raise G in order to overturn it. In Fig. 60, $W \times M\,N$. It is the "*work*" required to overturn it. The "*coefficient of stability*" $\times \frac{H\,A}{H\,R}$. Fig 60.

For double stability $\frac{H\,A}{H\,R} = 2$. When $\frac{H\,A}{H\,R} = \frac{9}{4} = 2\frac{1}{4}$, then $2\frac{1}{4}$ is the coefficient of stability. When R is at A, then $\frac{H\,A}{H\,R} = 1$ and we have simple stability. When R is at H, then $\frac{HA}{HR} = \frac{HA}{0} = \infty$; that is, no possible force can overturn the wall. If the resultant passes through the centre of the figure and of the base, the pressure at all parts would be uniform, and equal the whole pressure, divided by the area of the base. If it does not, the pressure increases on the side to which the resultant is nearest. It is found desirable in structures of masonry, resting on earth, that the maximum pressure should nowhere exceed double the mean pressure. To effect this for a rectangular base, make $\frac{HA}{HR} = 3$, which brings R at one-third the thickness from the front edge ; for a circular base make $\frac{HA}{HR} = 4$; for a hollow square, $1\frac{1}{2}$; for a circular ring, 2. These last two expressions apply to tall factory chimneys.

Masses resisting parallel forces as wind, like towers, chimneys, &c.

106. Any force, acting upon a convex surface, is two-thirds what it would be on a plane surface. Regarding the wind as a number of parallel forces, we consider the resultant at the centre of gravity.

We should examine whether the lower courses of bricks might not be crushed by the data given. The crushing force for common bricks is from 900 to 1,900 pounds per square inch, or 60 to 120 tons per square foot.

The force of waves acts similarly to that of winds ; but if the body be immersed in water, as in the case of the lower stones of a light house, the weight of the object will be diminished by that of an equal bulk of water.

Dams, Walls, &c.

107. It is cheaper to make the wall stable by adding to its thickness than to its height, for the reason that its breadth enters twice as a factor in the formula, 1st in weight, 2d in moment of stability. Hence a wall twice as thick is four times as strong.

To find a general expression for a rectangular wall which shall just stand while resisting a fluid. Let w = weight per cubic foot of fluid, w_1 = weight per cubic foot of wall, h = height of wall, t = thickness, l = length = 1. The pressure of the fluid $= h\frac{h}{2} \times 1 \times w$, and the moment of the fluid $= h \times \frac{h}{2} \times w \times \frac{h}{3} = \frac{w\,h^3}{6}$. The weight of wall $= h\,t \times 1 \times w_1$, and the moment of the wall $= h\;t \times 1 \times w_1 \times \frac{t}{2} = w_1\,h\frac{t^2}{2}$. For equilibrium; $\frac{w\,h^3}{6} = w_1\,h \times \frac{t^2}{2}$. Hence, $t = h\,\sqrt{\left(\frac{w}{3\,w_1}\right)} = \frac{4}{7}\,h\,\sqrt{\left(\frac{w}{w_1}\right)}$. In order to secure double stability $w_1\;h\;\frac{t^2}{2} = 2\,\frac{w\,h^3}{6}$, whence $t = h\,\sqrt{\left(\frac{2\,w}{3\,w_1}\right)}$ and this for water becomes $t =$ $.816\;h\,\sqrt{\left[\frac{62.5}{w_1}\right]} = \frac{6.45\;h}{\sqrt{w_1}}$.

Masses resisting earth, as retaining walls.

108. To find an expression for the thickness of a wall, resisting earth. In the expression for double stability against water, substitute for w the weight of earth $\times$ tan.^{2} $\frac{1}{2}$ the angle of the natural slope of the earth with the vertical (as in Part I). Let α = natural slope of the earth with the vertical. Then we have;

$$t = h \sqrt{\left(\frac{\frac{2}{3} w \times \tan.^2 \frac{1}{2} \alpha}{w_1}\right)} = .8 \times \tan. \tfrac{1}{2} \alpha \times h \sqrt{\left(\frac{w}{w_1}\right)}$$

Walls with sloping faces or batters.

A wall with vertical back and sloping face, will resist overturning with nearly twice the moment of a rectangular one of the same material. To find the bottom thickness of a wall with a triangular cross-section, which shall have the same stability as a given rectangular wall. Let h = height of wall. x = bottom thickness, t = thickness of rectangular wall, w^1 = weight per cubic foot. Moment of rectangular wall $= h\ t \times 1 \times w^1 \frac{t}{2} = \frac{1}{2}\ h\ t^2\ w^1$.

Moment of triangular wall $= \frac{h}{2}\ x \times 1 \times w^1\ \frac{2}{3}\ x = \frac{1}{3}\ h\ x^2\ w^1$. Hence, $x = 1.22\ t$, and the mean thickness of such a triangular wall $= .61\ t$, thus saving about .4 of the material. If the slope of the wall was on the inside, the wall would have only one-half its former stability. In practice a triangular wall is never built, for its top must have some thickness, and not less than two feet. The section of the wall then becomes a trapezoid.

To find the thickness of battering walls with the same stability as rectangular walls of the same height; (See R. & R. R., page 184,) subtract from the thickness of latter four-tenths of the entire projection of the batter. The remainder is the mean thickness of the required wall.

Approximately. At one-ninth of the height of the rectangular wall from the bottom, draw through its front a line having the desired batter. It will be the desired wall near enough for the usual batter.

Fig. 63.

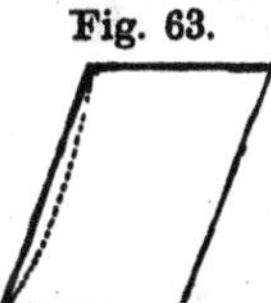

The final form, for economy of material, is shown in Fig. 63, the rear of the wall also sloping. The face is sometimes curved concave outwardly, as shown by the dotted line.

Walls supported by Counterforts or Buttresses.

109. These should be on the front of the wall so as to throw the centre of gravity farther from the edge of rotation, but for other reasons, are generally placed on the back of the wall. In calculating the moments of these counterforts, we may conceive them to be extended sidewise so as to fill the space between them and diminish proportionally in weight, and then reason about them as about a continuous wall.

Walls of Uncemented Blocks.

110. Let several stones be laid without cement, one upon another, and let a given pressure, P, act upon the uppermost stone at some point a. The weight of the stone acts vertically. At the point a construct a paral-

lelogram of these two forces whose diagonal will meet the second stone in some point b. On this diagonal, and the vertical at b, construct another parallelogram whose diagonal will meet the third stone in c; and so on for all the points, until a diagonal is found whose direction does not differ sensibly from the vertical, or direction of gravity. By these means we find the respective pressures at $a\,b\,c$, &c. The curved line joining these points is called the "*line of resistance.*" This line must intersect the common surface of each two contiguous portions of the structure within its mass.

Equation of the Line of Resistance.

Fig. 64.

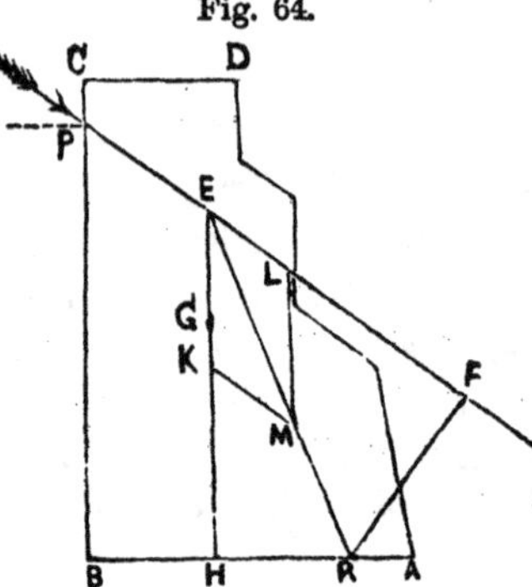

Let a force p act at a point P, Fig. 64, in a line $P\,F$, making an angle φ with the horizontal. Let G be the centre of gravity of the part of the buttress above $A\,B$. Compound the two forces and let their resultant pierce the joint $A\,B$ in R. Required the value of $R\,B$, for any value of $P\,B$. Let $P\,B = x$, $R\,B = y$, $B\,H = y'$, and W = wt. of buttress. The moment of the buttress $= W \times H\,R = W \times (y - y')$. The arm of the force $p =$ the perpendicular distance $R\,F$ of R from the line of action of the force p, $= x \times \text{cos. } \varphi - y \times \text{sin. } \varphi$. Hence the moment of $p = p\,(x \times \text{cos. } \varphi - y \times \text{sin. } \varphi)$. Equating,

$p\,(x \times \cos.\,\varphi - y, \sin.\,\varphi) = W\,(y - y')$. Hence,

$$y = \frac{W\,y' + p\,x\cos.\,\varphi}{W + p\sin.\,\varphi}.$$

When the force acts horizontal $\varphi = 0$, and the formula becomes

$$y = \frac{W\,y' + p\,x}{W} = y' + \frac{p\,x}{W}.$$

When the buttress is rectangular.

Let t = thickness, $y' = \frac{1}{2}\,t$, h' = height, above P, and h = total height. $W = w\,h\,t$, in which w = weight per cubic foot. The equation of the line of resistance becomes

$$y = \frac{\frac{1}{2}\,w\,(h' + x)\,t^2 + p\,x\cos.\,\varphi}{w\,(h' + x)\,t + p\sin.\,\varphi}.$$

This line is a rectangular hyperbola passing through the point E, and having a vertical asymptote whose distance from the face of the buttress $= \frac{1}{2}\,t + \frac{p\cos.\,\Phi}{w\,p}$, The

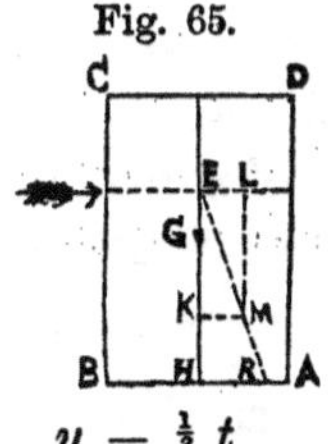

Fig. 65.

same result for a rectangular buttress and horizontal force may be obtained directly. From the similar triangles, Fig. 65, $E\,K\,M$ and $E\,H\,R$ we have $\frac{R\,H}{H\,E} = \frac{K\,M}{K\,E}$ or

$\frac{y - \frac{1}{2}\,t}{x} = \frac{p}{W} = \frac{p}{w\,t\,x + w\,t\,h'}$. Hence, $y = \frac{1}{2}\,t + \frac{p\,x}{w\,t\,x + w\,t\,h'}$. When $h' = 0$ we have $y = \frac{1}{2}\,t + \frac{p}{w\,t}$. That is, the line of resistance is then parallel to the face of the wall, and coincides with the asymptote.

ARCHES.

111. An arch is a structure of wedge shaped blocks, the extreme ones of which are confined between two supports. It is usually curved in one direction and is then used to resist a pressure against its convex side. Its most common application is to cover a space as a room, or to serve as a bridge. In them it has to support its own weight in addition to what other load may come on it. An arch may fail to do this in one of three ways, viz.: 1st. Its material may be crushed. 2d. Some portion of it may turn over upon others. 3d. Some portions of it may slide upon others.

It is therefore first necessary to investigate the amount and direction of the forces which are acting in any arch. Knowing this, we can determine by Part II. whether the materials will crush. Then by the preceding formulas in Part III., whether any portion will turn over, and finally, by Part IV., whether any portion will slide. The condition of an arch, neglecting the adhesion of mortar, is analogous to that of a wall of uncemented blocks, Art. 105. We have, consequently, to find the line of resistance, and then to determine its proper position, so that the arch shall be stable.

To determine the line of resistance in an arch.

112. *Coulomb's method.* This method determines the line at its extreme limit, when the arch is just about to give way by overturning. An arch may do this in three ways. 1st. In ordinary cases it yields thus: It opens at the crown on the intrados, on each side of the crown

at the extrados, and at the spring at the intrados, as shown in Fig. 66. The points D and D' are called "*points of rupture.*" At the moment at which the equilibrium is destroyed, the four parts into which the arch divides, touch each other (supposing them incompressible) only at the points A, D and D', and the ground only at B and B'. If then we conceive these points joined by straight lines, and the weights of the four parts of the arch applied at the points where the verticals passing through their centres of gravity meet these lines, then these lines represent the whole arch. We consider only a slice of the arch equal to a unit of length. Call p = weight of one of the upper pieces, q = weight of one of the lower ones, and G and G' their centres of gravity. Let $D\,K = x$, $B\,T = x'$, $B\,S = x''$, $A\,L = D\,H = y$, $B\,E = y'$, and $B\,Z = h$. Since the two parts of the system are symmetrical on each side of A, we may replace one of these by a horizontal force t, acting along $A\,Z$. We shall then have to consider only two bodies in equilibrium around the point B. Then the conditions of equilibrium, so that, there shall not be rotation in A, are $t\,y = p\,x$. That there may not be rotation at B, we have $t\,h = p\,(x' + x) + q\,x''$. But equilibrium is not sufficient; there must be an excess of stability, and we must have

Fig. 66.

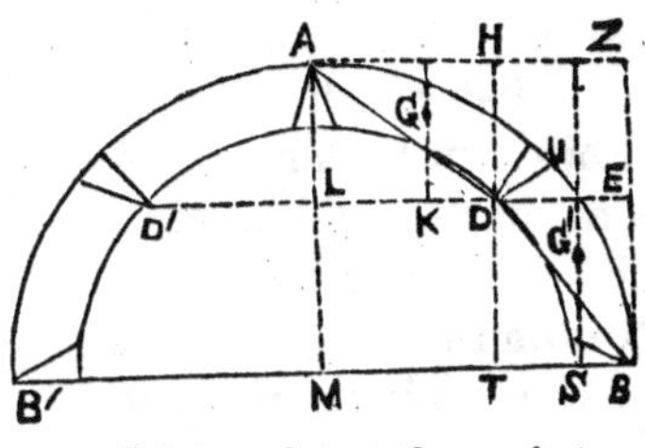

$$p\,(x' + x) + q\,x'' \text{ greater than } (t \times h) = \frac{p\,x}{y} \times h.$$

The first member of this inquality is determined by the form of the arch. The second will vary with the position of the point D. Among all possible values of the second member, the greatest one is that which concerns us, since it threatens most the stability of the arch. We have therefore to seek its maximum, since it is that which tends most to overturn the arch. In other words; to determine whether the arch will stand or fall, we must get the maximum moment of this thrusting force with respect to an extreme edge such as B, and also the moment of the entire semi-arch about that edge. If the latter moment be the greater, the arch will stand, and the greater its excess the greater will be its stability. The result of an analytical investigation is that the tangent to the curve of the intrados at the point of rupture, meets the horizontal drawn through the extrados at the key, in the vertical line passing through the centre of gravity of the part of the arch acting. Therefore to find the point of rupture, assume a point D for it, and at this point draw a tangent to the intrados meeting the horizontal line drawn through the top of the extrados.

Find the centre of gravity of the part of the arch between the point D and the crown. If the vertical passing through it and the other two lines meet in one point O the point has been rightly taken. If not try again. The difficulty is the laborious operation of finding the centre of gravity. It may be found in two ways, viz; by trial, and by an approximate rule.

By Trial. Cut the exact figure of the part under consideration from a piece of paste-board, and find its centre of gravity by trial.

Fig. 67.

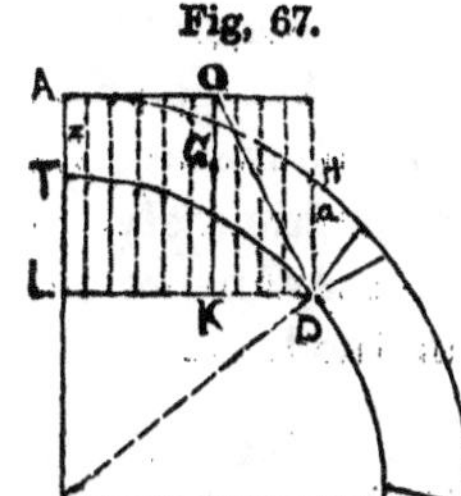

By an Approximate Rule. Divide $D\ L$ into n number of equal parts. Through the points of division, draw verticals to the top of the arch. Call the extreme ordinates a and z, and the other vertical distances between the intrados and extrados, (i. e. the top of load) $b.\ c.\ d.\ e.\ c.$, &c. Let the common distance between the verticals $= \delta$. Then DK

$$= \tfrac{1}{3}\,\delta \times \frac{a + (3\,n - 1)\,z + 6\,(b + 2\,c + 3\,d + 4\,e, \text{ \&c.})}{[a + z + 2\,(b + c + d +, \text{ \&c.})]}$$

If that part of the divisor in brackets be multiplied by $\frac{1}{2}\,\delta$, it will give the area of the portion $D\ T\ A\ H'$, H' being the highest point of the load or $D\ T\ A\ H$ when H is the highest point.

If the filling above the arch be lighter than the arch, its depth in our calculation for the centre of gravity, must be proportionally less. For example, if it be of earth of ½ the specific gravity of the masonry, consider it to be ½ as high as it really is. If the road way be supported by walls with empty spaces between them and they occupy ¼ the width of the bridges, call the filling ¼ as high as it is. In semi-circular arches the point of rupture, occurs from 25° to 31° from the spring. In an arch with a span of 65½ feet., depth of key stone 3¼, with a level roadway on top, the angle of rupture was 27°. In arches formed by segments of circles, in which the ratio of the rise to the span is less than .29 (which is the ratio of the versed sine to the chord of rupture in semi-circular arches) the point of rupture will be at the spring of the arch. In basket-handle arches the point of rupture is generally so situated that the normal to the intrados at that point makes an angle of 45° with the vertical. The above data gives the position of the point of rupture approximately.

The portion of the arch below this point may be regarded as part of the pier rather than of the arch.

Second manner of giving way.

Fig. 68.

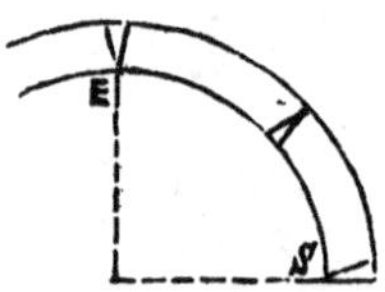

By rising at the key, opening at the extrados at the key, and on the intrados, at the haunches, and at the outside base of the piers, Fig. 68. In this case we have to find the minimum moment of the horizontal force at *E*, with respect to various interior edges, such as S, and the moment of the whole semi arch, about the same interior edges. If the former exceed the latter the arch will stand, and *vice versa*. The investigation in this case, resembles that of the first, except that the points of rotation are changed. This occurs very rarely, and only when the lower parts of the arch, that is its haunches, are excessively loaded.

Third manner of arches giving way. This is by an arch sinking at the key, and the pieces sliding on their bases. This will be examined in Part IV.

Mery's Graphic Method.

Fig. 69.

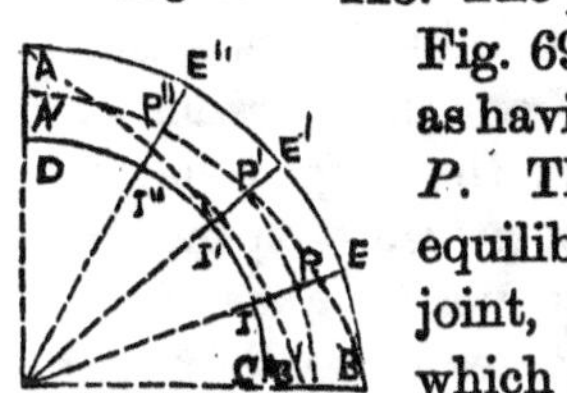

113. The pressures acting on any joint $I\ E$, Fig. 69, of an arch, may be considered as having one resultant in some point P. The weight of the arch is kept in equilibrium by the reaction p of the joint, and by the horizontal thurst t, which acts at the summit of the arch. If similar points P' P'', &c., be obtained for all joints, the curve formed by connecting them will be the line of resistance. The line of resistance is determined as in the case of a wall. (See. art., 110.)

We start with the horizontal thrust as in Coulomb's method, compounding this with the weight of the part of the arch above the joint in question, we obtain a resultant which will meet the joint in a point which will be in the line of resistance. So for all other joints. This line shows in what points and how the arch is in danger of giving way. Thus, if the line of resistance be $A\ I\ B'$, the arch tends to open at D, B, and E. If the line of resistance be $A'\ I\ B$, the arch tends to open in A, E, and C. It is desirable that the line should pass as nearly as possible through the middle of the arch, so that the pressure may be nearly equally distributed over the surface of the bed of each voussoir.

This is not absolutely necessary, but the following condition must be fulfilled: To prevent the material of an arch from being crushed, the line of resistance must always pass so far from the nearest surface of the arch, whether intrados or extrados, that the portion of the stone within this distance shall be able to support safely two-thirds of the total pressure on the whole surface of the stone. This takes place when the line passes within one-third the thickness of the arch from the intrados or extrados. The extreme pressure on the edge which it approaches, is then double the mean pressure on the entire bed, the other edge receiving none. We may then conceive the arch to be divided into three zones, the outer two of which shall satisfy the above conditions, and the middle one of which shall contain the line of resistance whatever variations in it may be caused by accidental loads or other pressures. In attempting to obtain this line, we meet with this difficulty. We do not know at what point of the key the horizontal thrust is applied. We must therefore try several. In a heavily proportioned arch, likely to open in the first mode, take the starting point above the middle of the key, for the line of pressure will then be there. It should not be taken nearer the top than one-third the depth, for the reasons just given.

In a light arch, so loaded as to be in danger of rising up at the crown, as in the second mode, and for a pointed arch, take the starting point at one-third the depth from the bottom of the keystone. If the lines thus obtained come too near the edge, try others until you find one that does come within the required conditions; or find that the arch does not contain any such line, which indicates that such an arch would not be safe. The angle which the line of pressure makes with the joints will determine whether one stone will slide on the other. (See table,

Part IV.) This, however, rarely happens and will be considered in Part IV.

A skew arch may be calculated by Mery's method, as if it were a right arch with a section equal to that parallel to the heads. The line of resistance should, however, come farther from the edge of the stones than in a right arch, in the ratio of 1 : cos. angle of skew; or sec. angle of skew : 1.

To determine the best form for the intrados of an arch, the span and rise being given.

The general principle is that the intrados should be parallel to the linear arch, which would correspond to the particular loading of the arch. Its own weight, must however, also be taken into account. When an arch has to support only its own weight, as when it is only a roof, its curve should be a catenary. Then the line of resistance on it would be a catenary, and everywhere parallel to the curve of the arch, and would therefore, pass through its middle. When the arch sustains a load, uniformly distributed horizontally. Its curve should be parabolic. For since its load is uniformly distributed, its resultant bisects $A\ x$; therefore, the tangent bisects $A\ x$; since these three forces acting, must meet in one point, therefore, the curve is a parabola. When the load acts vertically, and is distributed in any manner. Then the curve of equilibration is obtained thus: On a board, let a chain hang down, so as to have the requisite span and rise. This would be a catenary, and would be the proper curve for an unloaded arch. But when the arch is

Fig. 70.

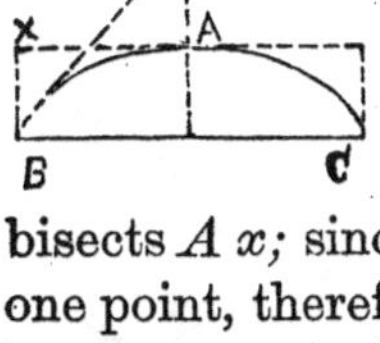

supporting a level roadway the weight will press more upon one part than upon another, hence, suspend weights or pieces of chains from the large chain, and terminate them as far below the curve as the roadway is to be above. This must be still farther modified when the arch and roadway are of different specific gravities. We must then so adjust the relative weights of the chain and the pieces hanging from it, that they shall be to each other in the same ratio as the stones of the arch bear to the load above them. An arch satisfying the above condition is in a state of perfect equilibrium; no one part having a tendency to yield before any other part. If in such an arch the joints are perpendicular to the intrados then the line of pressure of the voussoirs, will be perpendicular to every point. An arch of equilibration with horizontal roadway, is found by calculation to have the following curve, suppose the arch to have 100 feet span and 40 feet rise.

Abscissas from centre.	Corresponding ordinates.
0	6
2	6.03
4	6.14
10	6.91
15	8.12
20	9.93
25	12.49
30	15.89
35	20.07
40	26.89
45	35.13
50	46.00

For general purposes a segmental arch is to be preferred. The thrust of a segmental arch is the same as the thrust of a semicircular arch of the same radius, when the segment is not less than the arc between the points of rupture, and nearly so when less. For a moderate span an arc of 60°, or one in which the span equals the radius, is much used in France. Its rise is two-fifteenths span. Another favorite French form is one in which the rise is one-third the span. They call it "Surbaisse au tiers."

In basket-handle arches the rise should be between one-third and one-fourth span, approaching the former for small spans and the latter for large spans. Semicircular arches are very common, but not very advisable.

To determine the extrados of an arch, the intrados being given.

Thickness of crown, or depth of keystone.

115. Thus far the thickness of the arch has been assumed to be known. The depth at the keystone is usually at first determined empyrically.

Perronet's formula is, .035 span + 1 foot = $\frac{1}{29}$ span + 1 foot. This gives too much for great arches.

Dejardin's formula. For semicircular arches, r being the radius of intrados, depth at key = .1 r + 1 foot. For a segmental arch, comprising 60°, this $= \frac{5}{100} r + 1$ foot, $= \frac{1}{20} r + 1$ foot. For a segemental arch of 40°, this $= \frac{2}{100} r + 1$ foot $= \frac{1}{50} r + 1$ foot. For equilateral pointed arches (r = span) thickness $= \frac{5}{100} r + 1$ foot, which is the same as for a segmental arch of 60°.

Ellet's formula is, thickness at crown $= \frac{3}{8} \sqrt{\text{span}}$.

Trautwine proposes to use for thickness at the crown

for arches of the best cut stone, $\frac{36}{100}$ √ radius at crown. For second class work, $\frac{4}{10}$ √ rad. at crown. For rubble, or brick work, $\frac{45}{100}$ √ rad. at crown.

Barlow's formula. He would make the depth, in a somicircular arch = $\frac{1}{9}$ r. The material will greatly effect the decision as to thickness of keystone.

Rankine's formula, for depth of keystone in feet is, $\frac{35}{100}$ √ rad. at crown, for a single arch; or, $\frac{41}{100}$ √ rad. at crown for a series of arches. For example, the Dora Bridge, 160′ radius depth at crown, by calculation 4.38; actual depth 4.9 feet. The Chester Bridge, 140′ radius, depth by calculation 4.1 feet, actual depth 4 feet.

Table of depth of key, actually and by calculation.

Name of Bridge.	Span.	Rise.	Actual depth of key.	Depth by Perronet's formula.	By Ellet's formula.	By Dejardin's formula.
Neuilly, ...	127	31 1-2	5.2	5.3	4.1	7.3
Chester*, ...	200	42	4.	7.9	5.3	11.
Dora, ...	147	18 1-4	4.9	6.1	4.5	8.3
Victoria, ...	160	70	4.6	6.5	4.7	9.
Maiden-head, †	128	24	5.2	5.4	4.2	7.4
Pont de Sorelle	68	8.2	4.4	3.27	3.1	4.4
Telford's, ...	10	3 1-2	1.25	1.34	1.18	1.5
Do.	18	6	1.5	1.6	1.6	1.9
Do.	30	12	2,	2.	2.	2.5
Do, ...	50	15	2.5	2.7	2.6	3,5

* Segmental. † Elliptical.

Thickness of the arch at other points.

116. The pressure increases from crown to spring. The thickness should increase accordingly. An approximate

rule is this: the thickness of the arch at any joint should = its thickness at the key × sec. (or ÷ by cos.) of the angle which that joint makes with the vertical. The arch truly so called, does not extend below the joint which makes an angle of 60° with the vertical. The thickness at that point is double that at the key, since $\sec. 60° = \frac{1}{\cos. 60°} = 2$ The dimensions of the parts below that point, are determined as under the next head.

Rondelets graphic construction of the extrados.

Fig. 71.

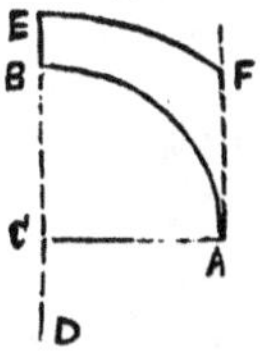

Through the spring A, draw an indefinite vertical line. Prolong the vertical radius $B\ C$, making $C\ D = \frac{1}{2}\ B\ C$. Then with a radius $D\ E$ describe an arc, cutting the vertical first drawn in F. $E\ F$ will be the extrados required. This construction is much used by architects in France. It, however, increases the volume of the arch and abutment, without increasing its stability. Yet is one of the best of the mere empyrical rules. Tables have been calculated, giving the thrust and proper dimensions of arches of the most usual forms. See Moseley's Mechanics of Engineers; and Captain Woodbury's Stability of Arches.

Dimensions of Piers.

117. Knowing the horizontal thrust of the arch, and the leverage, the stability can be determined by the principles of engineering statics. The horizontal thrust being supposed to be applied at the point of rupture.

The thickness thus obtained is that of mere equilibrium. It is usual to multiply the thickness thus obtained by a coefficient of from 1.25 to 1.40, which is equivalent to increasing the stability in a ratio = the square of these numbers, viz; 1.56 to 1.96. The following table gives suitable dimensions for semicircular arches. In this the spandrils are supposed to be filled up level and to have a level roadway, and over all a pavement sixteen inches thick.

Span.	Thickness at key.	Thickness at spring.	Thickness of abutments.	
			At ten feet high.	At twenty feet high.
5	1.25	1,4	2.6	2.9
10	1.40	1.7	3.5	4.2
15	1.60	2.1	4.1	4.8
20	1.80	2.5	4.8	5.8
25	1.95	3.0	5.5	6,6
30	2.10	3.5	6.2	7.5
40	2.50	4.8	7.6	9.1
50	2.80	5.9	8.6	10.4
100	4.50	10.7	13.5	16.0

The following table gives very safe dimensions:

Telford's Highland Bridges.

Span of arch.	Rise.	Depth of key stone.	Height of abutment.	Thickness of abutment.
4	1.6	1.0	2.6	1. 6
6	2.0	1.0	2.6	2. 0
8	3.0	1.2	2.6	2. 0
10	3.6	1.3	3.0	2. 6
12	4.0	1.4	3.0	3. 0
18	6.0	1.6	3.0	4. 6
24	8.0	1.9	4.0	5. 0
30	12.0	2.0	4·0	5. 6
50	15.0	2.6	6.0	6. 6

118. *Brick segmental arches.*

Span.	Rise.	Depth of key.	Thickness of abutment at top.
10	3.	1.2	3.0
20	6.	1.5	3.
30	9.	1.9	4.2
40	11.25	2.25	5.6

Abutment batter one-fourth inch to one foot.

For piers of bridges which have an arch on each side, and hence do not sustain any thrust, the usual thickness is two and one-half times the depth of the key stone of the arch.

The comparative thrusts of various forms of arches are in the following table :

	Pointed arches.	Semi-circular.	Arch lowered to 1-3.	Arch lowered to 1-6.	Arch lowered to 1-10.	Plate band.
Thrusts, ...	.49	1.	1.39	1.82	1.91	1.95
Thickness of piers, ...	.70	1.	1.18	1.35	1.39	1.42

The thickness of the piers varies as the square root of the thrust, since the thickness enters twice as a factor of their resistance.

PART IV.

STABILITY OF FRICTION, OR RESISTANCE TO SLIDING.

General Laws of Friction.

1. It is directly proportioned to the pressure.
2. It is independent of the extent of the surface.
3. It depends on the kind of surface.
4. It is independent of the velocity.

Recent experiments indicate, that it increases faster than the pressure, when that is so great as to produce abrasion, and that it increases somewhat with the surface.

Coefficient of Friction.

120. This is that fractional part of the pressure which resists motion, and which is expressed by f.

Limiting Angle of Resistance.

121. When one body rests on another, any force however small, acting on the former, will cause it to slide, if the direction of this force makes an angle with the perpendicular to the surface of contact, greater than a certain angle, depending on the nature of the surface. If the direction of the force makes an angle with the perpendicular to the surface, less than this same angle, the body will not slide however great the force. In most practical cases the direction of the effective force will be the resultant of the externally applied force and the weight of the body.

The "*coefficient of friction*" is the tangent of the limiting angle of resistance. Let P, Fig. 72, be a force acting on any particle W. Decompose $P\ W$ into $A\ W$ and $W\ V$. Then $W\ V$ denotes the pressure on the plane, and $A\ W$ the force tending to move particle along. Let the angle $P\ W\ V = a$. Then the pressure on the plane $W\ V = P$ cos. a. Resistance of friction $= f \times P$ cos. a. Pressure tending to move the particle $= A\ W = P$ sin. a. When motion is just ready to ensue, these must be equal, that is $f \times P$ cos. $a = P$ sin. a. Whence, $f = \frac{P \text{ sin. } a}{P \text{ cos. } a} = \text{tan. } a$. The "limiting angle of resistance" we will call φ. We will then have $f = \text{tan. } \varphi$. The cone of resistance is a conical surface, generated by a line passing around the perpendicular to the surface of contact, and making with it an angle equal to the limiting angle of resistance.

Fig. 72-

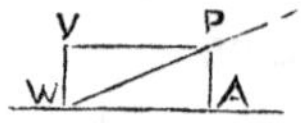

Angle of repose.

122. If a body rests on an inclined plane, the angle which this plane makes with the horizon, when the body is just about to slide, is called the "angle of repose." It equals φ, or the limiting angle of resistance whose tan. $= f$. The angle of repose of earth is the same as the natural slope of the earth. The greater the friction, the greater is φ, of which it is the tangent, and hence the more may the plane be inclined before the body will slip on it.

Demonstration that the angle of Repose equals the limiting angle of resistance. Fig. 73.

Fig. 73.

$F\,G : F\,E :: B\,C : A\,C$ or $:: A\,B \times \sin. a : A\,B \times \cos. a$, or $:: \sin. a : \cos. a \; \therefore \; \frac{F\,G}{F\,E} = \frac{\sin.\ a}{\cos.\ a} = \tan. a$. When the body is just about to slide, $F\,G = f \times F\,E$, whence $\frac{F\,G}{F\,E} = f = \tan. a$; but f is also the tangent of the limiting angle of resistance, hence, that angle equals the angle of repose, i. e., $a = \varphi$.

Table of coefficients of friction and limiting angles of resistance.

	f.	*a.*
Dry masonry and brick work, ...	.6 to .7	31° to 35°
Masonry and brick work, cemented, ...	.74	36° 1-2
Timber on stone, ...	.4	22°
Iron on stone, ...	.7 to .3	35° to 16° 2-3
Timber on timber, ...	.5 to .2	26°1-2 to 11°1-3
Timber on metal, ...	.6 to .2	31° to 11° 1-3
Metal on metal, ...	.25 to .15	14° to 8° 1-2
Masonry on dry clay, ...	.51	27°
Masonry on moist clay, ...	.33	18° 1-4
Earth on earth. Dry sand and clay and mixed earth, ...	.38 to .75	21° to 37°
Earth on earth. Damp earth, ...	1.	45°
Earth on earth. Wet clay, ...	.31	17$_o$
Earth on earth. Shingle and gravel, ...	.81 to 1.11	39$_o$ to 48°

Structures of uncemented Blocks.

123. If the resultant, as obtained for any joint by Art. 110, makes an angle with the perpendicular to the surface of contact, greater than the limiting angle of resistance for that surface, then the wall will give way by that stone sliding on the one below it, however small the pressure of the resultant. The line of resistance is continually approaching the vertical, hence the tendency to slide is continually diminishing in descending through the structure.

Line of Pressure.

124. The line of resistance is obtained (as shown in Part III) by joining the points at which the joints meet the resultants of the externally applied force. Now let the resultants be produced indefinitely they will meet in a series of points. The curved line passing through these points is called by Moseley the "*Line of Pressure.*" All the resultants are tangent to it. Then to determine whether the structure will slide at any joint, draw from the point where the "line of resistance" intersects the joint, a tangent to the "line of pressure." The angle which this line makes with the joint, will determine whether the wall will slide at that joint or not, according as the angle be more or less than the "limiting angle of resistance." The whole theory of equilibrium of any structure, depends on these two lines. (See Woodbury on the Arch.)

The "line of resistance" determines the point of application of the resultant of the pressure on each surface of contact, and the "line of pressure" determines the direction of that resultant. The arch is only one form of the wall before mentioned.

Piles Supported by Friction.

125. Let R = resistance which the earth opposes to the sinking of a pile during its sinking; s = space the pile is sunken by one blow of the ram. Then neglecting the elasticity and assuming R constant during s, we have $\frac{w^2}{w+w^1} \times h = R\,s - (w + w^1)\,s$, whence, $R = \frac{w^2}{w+w^1} \times \frac{h}{s} + w + w^1$. Neglecting w and w^1 as very small when compared with R, and we have approximately

$$R = \frac{w^2}{w+w^1} \times \frac{h}{s} = \frac{w}{1+\frac{w^1}{w}} \times \frac{h}{s}.$$

Hence, we infer that the dead weight which a pile will bear is directly as h, and inversely as s, and directly as $\frac{w}{1+\frac{w^1}{w}}$ or nearly as w. For a permanent load from one-fifth to one-tenth of the value of R is used.

For another formula see Rankine's Applied Mechanics, p. 565. Airy's formula, modified by Rankine, is this: Let R, S, h, w, mean as before, let e = modulus of elasticity for the material of the pile. Tables of its value are given in Part II. Let a = area of the head of the pile, l = ½ its length

$$R = \sqrt{\left[\frac{2\,e\,a\,w\,h}{l} + \left(\frac{e\,a\,s}{l}\right)^2\right]} - \frac{e\,a\,s}{l}.$$

The factor of safety is from 3 to 10.

McAlpine's formula is: extreme supporting power $= P = 80\,(W + .228\,\sqrt{F} - 1)$. In which W = weight of ram in tons, and F = fall in feet.

TABLE OF CONTENTS.

PART I.

Forces to be resisted.

PART II.

Strength of Cohesion.

www.ingramcontent.com/pod-product-compliance
Lightning Source LLC
LaVergne TN
LVHW021409110826
845150LV00007B/1846